SOLUTIONS OF
ILL-POSED PROBLEMS

SCRIPTA SERIES IN MATHEMATICS

Tikhonov and Arsenin · Solutions of Ill-Posed Problems, 1977

Rozanov · Innovation Processes, 1977

SOLUTIONS OF ILL-POSED PROBLEMS

ANDREY N. TIKHONOV
Moscow State University

and

VASILIY Y. ARSENIN
Moscow Institute of Engineering Physics

Translation Editor

FRITZ JOHN
New York University–Courant
Institute of Mathematical Sciences

1977
V. H. WINSTON & SONS
Washington, D.C.

A HALSTED PRESS BOOK

JOHN WILEY & SONS
New York Toronto London Sydney

V. H. Winston & Sons, a Division of Scripta Technica, Inc., Publishers
1511 K Street, N.W., Washington, D.C. 20005

Distributed solely by Halsted Press, a Division of John Wiley & Sons, Inc.

Library of Congress Cataloging in Publication Data

Tikhonov, Andreĭ Nikolaevich, 1906-
Solutions of ill-posed problems.

(Scripta series in mathematics)
An English-language translation of Методы решения некорректных задач, published by Наука in the USSR.
Bibliography: p.
Includes index.
1. Numerical analysis—Improperly posed problems.
I. Arsenin, Vasiliĭ IAkovlevich, joint author. II. Title.
III. Series.
QA297.T5413 1977 519.4 77-3422
ISBN 0-470-99124-0

Composition by Marie A. Maddalena, Scripta Technica, Inc.

CONTENTS

TRANSLATION EDITOR'S PREFACE

The notion of a well-posed ("correctly set") mathematical problem makes its debut with the discussions in Chapter I of J. Hadamard's Lectures on Cauchy's Problem in Linear Partial Differential Equations (Yale University Press, 1923). It represented a significant step forward in the classification of the multitude of problems associated with differential equations, singling out those with sufficiently general properties of existence, uniqueness and (by implication) stability of solutions. Hadamard observes:

"But it is remarkable, on the other hand, that a sure guide is found in physical interpretation: an analytic problem always being correctly set, in our use of the phrase, when it is the translation of some mechanical or physical question."

From the point of view of the present volume, Hadamard's notion of a mechanical or physical question appears as too narrow. It applies when the problem is that of determining the effects ("solutions") of a complete set of independent causes ("data"). But in many applied problems we have to get along without a precise knowledge of the causes, and in others we are really trying to find "causes" that will produce a desired effect. We are then led to "ill-posed" problems. One might say that the majority of applied problems are, and always have been, ill-posed, particularly when they require numerical answers.

The general methods of mathematical analysis were best adapted to the solution of well-posed problems; moreover, it was not clear in what sense ill-posed problems could have solutions that would be meaningful in applications. A. N. Tikhonov was one of the earliest workers in the field of ill-posed problems who succeeded in giving a precise mathematical definition of "approximate solutions" for general classes of such problems, and in constructing "optimal" solutions. The book by A. N. Tikhonov and V. Ya. Arsenin gives a general theory of ill-posed problems and introduces the reader to a wide variety of applications ranging from heat flow to the design of optical systems and the automatic processing of observational data. The present translation should be of special interest to mathematicians and scientists concerned with the numerical solution of applied problems, but are unfamiliar with Russian language literature on the subject. But not only to those. Past experience suggests that the concepts and methods used in the discussion of ill-posed problems will in turn stimulate advances in "pure" mathematical analysis.

Fritz John

PREFACE TO ENGLISH EDITION

The advent of the computer had forced the application of mathematics to all branches of human endeavor. This broadening of the range of influence of mathematics is one of the most import signs of scientific and technological progress, and has produced a need to take a new look at the potential applications of this science.

One important property of mathematical problems is the stability of their solutions to small changes in the initial data. Problems that fail to satisfy this stability condition are called, following Hadamard, *ill-posed.** For a long time mathematicians felt that ill-posed problems cannot describe real phenomena and objects. However, we shall show in the present book that the class of ill-posed problems includes many classical mathematical problems and, most significantly, that such problems have important applications.

If the initial data in such problems are known only approximately and contain a random error, then the above-mentioned instability of their solutions leads to nonuniqueness of the classically-derived approximate solutions and to serious difficulties in their physical interpretation. Also, in many cases

---

*Also called *improperly posed* or *incorrectly posed* by some authors (Editor).

there simply is no classical solution of problems with approximate initial data. In solving such problems, one must first define the concept of an approximate solution that is stable to small changes in the initial data, and to use special methods for deriving this solution.

Ill-posed problems include such classical problems of analysis and algebra as differentiation of functions known only approximately, solution of integral equations of the first kind, summation of Fourier series with approximate coefficients, analytical continuation of functions, finding the inverse Laplace transform, the Cauchy problem for the Laplace equation, solution of singular or ill-conditioned systems of linear algebraic equations, and many others, some of which are of great practical significance. We can divide all such problems into two subclasses, namely, those of recognition and those of design. The first subclass includes, among others, problems of mathematical processing and interpretation of data, as in nuclear-, plasma-, or radiophysics, electronics, interpretation of geophysical observations, mineral (including petroleum) prospecting, rocketry, nuclear reactor engineering, etc. The second subclass includes many optimization problems, for example, design of antenna or optical systems, optimal control, and optimal economic planning.

Our book deals with solutions of problems of this nature. It defines the concept of an approximate solution of such problems that is stable to small changes in the initial data and it examines in detail methods of constructing solutions that are easily processed on a computer.

As we have already said, the initial data underlying ill-posed problems (generally measurements) contain random errors. Depending on the nature of this initial information, one can take either a deterministic or a probabilistic approach to derivation of the approximate solutions or in evaluating the error of the data. We have generally (except for Chapters IV and V) confined ourselves to the deterministic approach. The interested reader will find the probabilistic approach discussed, among others, in references 63, 64, 87, 100, 105, 124, 128, 129, 179–181, 184 and 217.

One feature of the book is the development of the regularization method in construction of approximate solutions of ill-posed problems that was first expounded in [156–161].

We have not attempted a survey of the literature on ill-posed problems. Therefore, the bibliography at the end does not pretend to be complete. However, one recommended survey of the literature is that of Morozov [122].*

The book is designed for a broad range of scientists, teachers, engineers and students interested in mathematical processing of information and prediction of experiments.

We thank A. V. Lukshin for many valuable comments.

A. N. Tikhonov and V. Ya. Arsenin

---

*See also L. Payne: *Improperly Posed Problems* (to appear shortly) [Editor].

SIAM CBMS Series #22? 1975 or 1976

INTRODUCTION

§ 1. Remarks on the formulation of mathematical problems.

1. The rapidly increasing use of computational technology requires the development of computational algorithms for solving broad classes of problems. But just what do we mean by the "solution" of a problem? What requirements must the algorithms for finding a "solution" satisfy?

Many features of problems encountered in practice are not reflected by the classical conceptions and formulation of problems. Let us illustrate this with some examples.

2. Example 1. Consider the Fredholm integral equation of the first kind with kernel $K(x, s)$:

$$\int_a^b K(x, s) z(s) ds = u(x), \quad c \leqslant x \leqslant d, \tag{I.1.1}$$

where $z(s)$ is the unknown function in a space F and $u(x)$ is a known function in a space U. Let us assume that the kernel $K(x, s)$ is continuous with respect to the variable x and that it has a continuous partial derivative $\partial K/\partial x$. For brevity, we shall denote by A the operator defined by

$$Az = \int_a^b K(x, s) z(s) ds$$

We shall seek a solution $z(s)$ in the class C of functions that are continuous on the interval $[a, b]$. We shall measure changes in the right-hand member of the equation with the L_2-metric defined by

$$\rho_U(u_1, u_2) = \left\{ \int_c^d [u_1(x) - u_2(x)]^2 dx \right\}^{1/2}$$

while we measure changes in the solution $z(s)$ in the C-metric defined by

$$\rho_F(z_1, z_2) = \max_{s \in [a, b]} |z_1(s) - z_2(s)|$$

3. Suppose that for some $u = u_1(x)$ the function $z_1(s)$ is a solution of equation (I.1.1):

$$\int_a^b K(x, s) z_1(s) ds \equiv u_1(x).$$

If instead of the function $u_1(x)$ we know only an approximation $u(x)$ that differs slightly (in the L_2-metric) from $u_1(x)$, we can speak only of finding an approximate solution of equation (I.1.1), that is, a function close to $z_1(s)$.

Here, the right-hand member $u(x)$ can be obtained experimentally, for example, with the aid of a recording device, and it may have "corners" at which the function $u(x)$ does not have a derivative. With such a right-hand member, equation (I.1.1) does not have a solution (in the classical sense) since the kernel $K(x, s)$ has a continuous derivative with respect to x and hence the right-hand member must also have a continuous derivative with respect to x.

This means that we cannot take for the "solution" of equation (I.1.1) approximating $z_1(s)$ the exact solution of that equation with the approximately known right-hand member $u(x) \neq u_1(x)$ since such a solution may not exist. The fundamental question then arises as to what we should mean by an approximate solution of equation (I.1.1) with approximately known right-hand member.

Obviously, equation (I.1.1) has a solution (in the classical sense) only for right-hand members $u(x)$ that belong to the image AF of the set F of functions $z(s)$ under the mapping

$$u = Az \equiv \int_a^b K(x, s) z(s)\, ds, \quad z(s) \in F.$$

4. Furthermore, the solution of equation (I.1.1), understood in the classical sense, that is, obtained according to the rule

$$z = A^{-1} u,$$

where A^{-1} is the inverse of the operator A, does not possess the property of stability under small changes in the "initial data" (the right-hand member $u(x)$).

To see this, note that the function $z_2(s) = z_1(s) + N \sin \omega s$ is a solution of equation (I.1.1) with right-hand member

$$u_2(x) = u_1(x) + N \int_a^b K(x, s) \sin \omega s\, ds.$$

Obviously, for any number N, if the values of ω are sufficiently great, we can make the change

$$\rho_U(u_1, u_2) = |N| \left\{ \int_c^d \left[\int_a^b K(x, s) \sin \omega s\, ds \right]^2 dx \right\}^{1/2}$$

arbitrarily small without preventing the change in the corresponding solutions $z_1(s)$ and $z_2(s)$, namely,

$$\rho_F(z_1, z_2) = \max_{s \in [a\, b]} |z_2(s) - z_1(s)| = \max_{s \in [a, b]} |N \sin \omega s| = |N| \tag{I.1.2}$$

from the being arbitrarily great. Here, we measured the difference between the functions $z_1(s)$ and $z_2(s)$ in the C-metric.

If the difference between the solutions is estimated in the L_2-metric, the solution of equation (I.1.1) is again unstable under small changes in $u(x)$. Specifically,

$$\rho_F(z_1, z_2) = \left\{\int_a^b |z_1(s) - z_2(s)|^2 ds\right\}^{1/2} = |N| \left\{\int_a^b \sin^2 \omega s\, ds\right\}^{1/2} =$$

$$= |N| \sqrt{\frac{b-a}{2} - \frac{1}{2\omega} \sin \omega (b-a) \cos \omega (b+a)}\,. \tag{I.1.3}$$

One can easily see that the numbers ω and N can be chosen in such a way that, for arbitrarily small discrepancies between $u_1(x)$ and $u_2(x)$, the discrepancy between the corresponding solutions as calculated according to formula (I.1.3) can be arbitrary.

However, the requirement of stability of the solution of equation (I.1.1) has to be satisfied since it is associated with the physical determinacy of the phenomenon (relationship) described by that equation.

The problem of solving the integral equation (I.1.1) is not completely settled by taking as an approximate solution a function $z(s)$ for which

$$\rho_U(Az, u) \leqslant \delta,$$

where u approximates the right-hand member of the equation with accuracy δ: $\rho_U(u, u_1) \leqslant \delta$.

Thus, we need not only to answer the question as to what is meant by an approximate solution of equation (I.1.1) but also to give an algorithm for constructing one that will be stable under small changes in the initial data $u(x)$.

The situation encountered in this example is typical of ill-posed problems.

5. We have considered a case in which it is known in advance that equation (I.1.1) has an exact solution (in the classical sense) $z_T(s)$ corresponding to the right-hand member $u_T(x)$ and we are required to find an approximation for it if, instead of the function $u_T(x)$, we know only an approximation $u(x)$ of it such that $\rho_U(u_T, u) \leqslant \delta$.

If we have no information regarding the existence of an exact solution of equation (I.1.1) but do have information regarding the class of possible right-hand members U, we can also pose the problem of finding an approximate solution of equation (I.1.1). To do this, we define the concept of a generalized solution (quasisolution) of equation (I.1.1) on the set F as an element $\tilde{z}$ of F at which the distance $\rho_U(Az, u)$ attains its greatest lower bound [71, 72]; that is,

$$\rho_U(A\tilde{z}, u) = \inf_{z \in F} \rho_U(Az, u), \quad Az \equiv \int_a K(x, s) z(s)\, ds.$$

Obviously, if equation (I.1.1) has a solution $z_T \in F$ in the usual sense for $u = u_T$, it coincides with the generalized solution of the equation $Az = u_T$.

The problem then arises of finding algorithms for constructing generalized solutions that are stable under small changes in the right-hand member $u(x)$.

Example 2. Consider the system of linear algebraic equations

$$Az = u, \tag{I.1.4}$$

where z is an unknown vector, u is a known vector, and $A = \{a_{ij}\}$ is a square matrix with elements a_{ij}.

If the system (I.1.4) is nonsingular, that is, if $\det A \neq 0$, it has a unique solution, which we can find by Cramer's rule or by some other method.*

If the system (I.1.4) is singular, it will have a solution (not unique) only when the condition for existence of a solution (vanishing of the relevant determinants) is satisfied.

Thus, before solving the system (I.1.4), we need to check whether it is singular or not. To do this, we need to evaluate the determinant of the system $\det A$.

If n is the order of the system, evaluation of $\det A$ requires approximately n^3 operations. No matter how accurately we perform the calculations, the value that we obtain for $\det A$ will differ from the true value as a result of the accumulation of computational errors, and this discrepancy will be arbitrarily great for sufficiently large n. This circumstance necessitates construction of algorithms for solving the system (I.1.4) that do not involve preliminary determination as to whether the system is singular or not.

Furthermore, in practical problems, we often know only approximately the right-hand member u and the elements of the matrix A, that is, the coefficients in the system (I.1.4). In such cases, we are dealing not with the system (I.1.4) but with some other system $\tilde{A}z = \tilde{u}$ such that $\| \tilde{A} - A \| \leqslant \delta$ and $\| \tilde{u} - u \| \leqslant \delta$, where the meaning of the norm is usually determined by the nature of the problem. Having the matrix $\tilde{A}$ instead of the matrix A, we are even less able than before to draw a definite conclusion as to whether the system (I.1.4) is singular or not.

In such a case, all we know regarding the exact system $Az = u$ is that $\|\tilde{A} - A\| \leqslant \delta$ and $\|\tilde{u} - u\| \leqslant \delta$. But there are infinitely many systems with such initial data (A, u), and we cannot distinguish between them within the framework of the level of error known to us. Such possible exact systems may include singular ones.

---

*Kurosh, A. G., *Kurs vysshey algebry* (Course of higher algebra), 10th ed., Moscow, Nauka, 1971.

Since we have the approximate system $\tilde{A}z = \tilde{u}$ rather than the exact one (I.1.4), we can speak only of finding an approximate solution. But the approximate system may be unsolvable. The question then arises as to what we are to understand by an approximate solution of the system (I.1.4). It must also be stable under small changes in the initial data (A, u).

§ 2. The concepts of well-posed and ill-posed problems.

1. Well-posed and ill-posed problems are distinguished. The concept of a well-posed problem of mathematical physics was introduced by J. Hadamard [201, 202] in the attempt to clarify what types of boundary conditions are most natural for various types of differential equation (for example, the Dirichlet and analogous problems for elliptic equations and the Cauchy problem for hyperbolic equations).

The solution of any quantitative problem usually ends in finding the "solution" z from given "initial data" $u: z = R(u)$. We shall consider u and z as elements of metric spaces U and F with metrics $\rho_U(u_1, u_2)$ for $u_1, u_2 \in U$ and $\rho_F(z_1, z_2)$ for $z_1, z_2 \in F$. The metric is usually determined by the formulation of the problem.

2. Suppose that the concept of solution is defined and that to every element $u \in U$ there corresponds a unique solution $z = R(u)$ in the space F.

The problem of determining the solution $z = R(u)$ in the space F from the initial data $u \in U$ is said to be **stable on the spaces** (F, U) if, for every positive number ϵ, there exists a positive number $\delta(\epsilon)$ such that the inequality $\rho_U(u_1, u_2) \leqslant \delta(\epsilon)$ implies $\rho_F(z_1, z_2) \leqslant \epsilon$, where $z_1 = R(u_1)$ and $z_2 = R(u_2)$ with u_1 and u_2 in U and z_1 and z_2 in F.

The problem of determining the solution z in the space F from the "initial data" u in the space U is said to be **well-posed on the pair of metric spaces** (F, U) if the following three conditions are satisfied:

1) for every element $u \in U$ there exists a solution z in the space F;

2) the solution is unique;

3) the problem is stable on the spaces (F, U).

For a long time, it was the accepted point of view in mathematical literature that every mathematical problem has to satisfy these conditions [93].

Problems that do not satisfy them are said to be **ill-posed.**

It should be pointed out that the definition of an ill-posed problem is *with respect to a given pair of metric spaces* (F, U) since the same problem may be well-posed in other metrics (see examples 2 and 3 of §3 of this introduction).

Remark. The fact that the spaces F and U are metric spaces is used here to express the closeness of elements as a means of describing neighborhoods in the spaces F and U. The basic results to be given below remain valid for topological spaces F and U (see [73]).

If the class U of initial data is chosen in a "natural" manner for the problem in question, conditions 1) and 2) characterize its mathematical determinacy. Condition 3) is connected with the physical determinacy of the problem and also with the possibility of applying numerical methods to solve it on the basis of approximate initial data.

When we get right down to it, what physical interpretation can we give to the solution of a problem if arbitrarily small disturbances in the initial data can cause great changes in the solution?

3. That a problem be well-posed has often been treated as a condition that any mathematical problem corresponding to any physical or technological problem must satisfy. This generated doubt as to the usefulness of studying ill-posed problems. However, this point of view, while perfectly natural as applied to certain phenomena that have been studied over the years, is not valid for all problems. In §3, we shall give some examples of ill-posed problems relating both to the basic mathematical apparatus and also to a broad class of applied problems.

4. The problem of finding an approximate solution of an ill-posed problem in a natural class F is in practice ambiguous.

Thus, suppose that the problem of solving the equation

$$Az=u \tag{I.2.1}$$

in the class F is ill-posed.*

Let us suppose that the right-hand member of equation (I.2.1) is known to us with accuracy δ; that is, we know not its exact value u_T but an element $\tilde{u}$ such that $\rho_U(u_T, \tilde{u}) \leqslant \delta$.

It is natural to seek an approximate solution of equation (I.2.1) in the class Q_δ of elements z for which $\rho_U(Az, \tilde{u}) \leqslant \delta$.

However, in a number of cases, this class of elements is too broad. As was shown in example 1 of §1, these elements (functions $z(s)$) include elements that can differ greatly from each other (in the metric of the space F). Therefore, not every element of the class Q_δ can be taken as an approximate solution of equation (I.2.1).

A rule for selecting possible solutions is necessary. For this, we need to use supplementary information (generally available) concerning the solution. Such information may be of a quantitative or qualitative nature. The attempt to use supplementary information of a quantitative nature leads to the concept of a **quasisolution** [71, 72], of which we shall speak in greater detail in Chapter I.

The use of supplementary information of a qualitative nature (for example, smoothness of the solution) necessitates a different approach to the construction of approximate solutions of equation (I.2.1). This approach is followed in [156–161] and will be described in Chapter II.

§3. Examples of ill-posed problems.

1. Example 1. The problem of solving an integral equation of the first kind

$$\int_a^b K(x, s) z(s)\, ds = u(x).$$

---

*This will be the case, for example, if A is a completely continuous operator. Then, the inverse operator A^{-1} will not in general be continuous on U and the solution of equation (I.2.1) will not be stable under small changes in the right-hand member u (in the metric of the space U).

We looked at this example in detail in §1, where it was pointed out that its solution is unstable under small changes in the right-member $u(x)$.

Example 2. The problem of differentiating a function $u(t)$ that is known only approximately.

Suppose that $z_1(t)$ is the derivative of the function $u_1(t)$. The function $u_2(t) = u_1(t) + N \sin \omega t$ differs from $u_1(t)$ in the metric of C by an amount $\rho_C(u_1, u_2) = |N|$ for arbitrary values of ω. However, the derivative $z_2(t) = u_2'(t)$ differs from $z_1(t)$ in the C-metric by an amount $|N\omega|$, which can be arbitrarily great for sufficiently large values of $|\omega|$.

We note that the problem of finding the nth derivative of the function $u(t)$ reduces to solving the integral equation of the first kind

$$\int_0^t \frac{1}{(n-1)!} (t-\tau)^{n-1} z(\tau)\, d\tau = u(t).$$

Thus, this problem does not possess the property of stability, a fact that leads to great difficulties in approximate evaluation of derivatives.

Remark 1. If we take other metrics on the sets F and U (or on one of them), then the problem of differentiating $u(t)$, which is known only approximately, may be well-posed on the pair of metric spaces (F, U).

Thus, if U is the set of continuously differentiable functions on the interval $[a, b]$ and the distance between two functions $u_1(t)$ and $u_2(t)$ in U is measured in the metric defined by

$$\rho_U(u_1, u_2) = \sup_{t \in [a,b]} \{ |u_1(t) - u_2(t)| + |u_1'(t) - u_2'(t)| \},$$

but the distance between two functions $z_1(t)$ and $z_2(t)$ in F is measured in the C-metric, then the problem of differentiation obviously is well posed in that pair of metric spaces (F, U).

However, in practical problems, these requirements on the functions $u(t)$ often cannot be verified. Therefore, the metric used above for estimating the difference between functions $u(t)$ in U is not a natural one for the differentiation problem.

Example 3. Numerical summation of Fourier series when the coefficients are known approximately in the metric of l_2.

Suppose that $f_1(t) = \sum_{n=0}^{\infty} a_n \cos nt$. If instead of a_n we take the coefficients $c_n = a_n + \epsilon/n$ for $n \geqslant 1$ and $c_0 = a_0$, we obtain the series $f_2(t) = \sum_{n=0}^{\infty} c_n \cos nt$. The coefficients in these series differ (in the metric of l_2) by an amount

$$\epsilon_1 = \left\{\sum_{n=0}^{\infty} (c_n - a_n)^2\right\}^{1/2} = \epsilon \left\{\sum_{n=1}^{\infty} \frac{1}{n^2}\right\}^{1/2} = \epsilon \sqrt{\frac{\pi^2}{6}},$$

which we can make arbitrarily small by choosing ϵ sufficiently small. At the same time, the difference

$$f_2(t) - f_1(t) = \epsilon \sum_{n=1}^{\infty} \frac{1}{n} \cos nt$$

may be arbitrarily large (for $t = 0$, the last series diverges).

Thus, if we take the deviation of the sum of the series in the metric of C, summation of the Fourier series is not stable.

Remark 2. If the difference between functions $f(t)$ in F is estimated in the metric of L_2, the problem of summation of Fourier series with coefficients given approximately (in the metric of l_2) will be well-posed on such a pair of metric spaces (F, U). Specifically, from Parseval's theorem, we have

$$\left\{\int_0^\pi [f_1(t) - f_2(t)]^2\,dt\right\}^{1/2} = \left\{\sum_{n=1}^\infty \frac{\pi}{2}(c_n - a_n)^2\right\}^{1/2} = \epsilon_1\sqrt{\frac{\pi}{2}}.$$

Example 4. The Cauchy problem for the two-dimensional Laplace's equation (Hadamard's example [202]). This problem consists in finding a solution of the equation $\Delta u(x, y) = 0$ from the initial data, that is, in finding a solution satisfying the conditions

$$u(x, 0) = f(x), \quad \left.\frac{\partial u}{\partial y}\right|_{y=0} = \varphi(x), \quad -\infty < x < \infty,$$

where $f(x)$ and $\varphi(x)$ are given functions.

If we set $f_1(x) \equiv 0$ and $\varphi_1(x) = a^{-1} \sin ax$, then the solution of the Cauchy problem will be the function $u_1(x, y) = a^{-2} \sin ax \sinh ay$, where $a > 0$.

If we set $f_2(x = \varphi_2(x) \equiv 0$, then the solution of this Cauchy problem is $u_2(x, y) \equiv 0$.

If we estimate the differences in initial data and solutions in the metric of C, we have

$$\rho_C(f_1, f_2) = \sup_x |f_1(x) - f_2(x)| = 0,$$

$$\rho_C(\varphi_1, \varphi_2) = \sup_x |\varphi_1(x) - \varphi_2(x)| = \frac{1}{a}.$$

The second of these can be made arbitrarily small by taking a sufficiently large. However, for any fixed $y > 0$, the difference between the solutions

$$\rho_C(u_1, u_2) = \sup_x |u_1(x, y) - u_2(x, y)| =$$

$$= \sup_x \left|\frac{1}{a^2} \sin ax \sinh ay\right| = \frac{1}{a^2} \sinh ay$$

can be made arbitrarily large for sufficiently large values of a.

Thus, this problem is not stable and hence is ill-posed.

However, the Cauchy problem for Laplace's equation is encountered in applications [68–70, 75, 84, 94, 96, 209, 211, 212]. As an example, we may cite the problem of continuation of the gravitational potential observed on the surface of the earth ($y = 0$) in the direction away from the gravitational field sources.*

Example 5. The problem of analytic continuation of a function known on part of a region to the entire region.

Let D denote a bounded region and let E denote an arc of a curve contained in D. Then, the problem of analytic continuation of the function represented by the arc E to the entire region D is unstable.

To see this, let z_0 denote a point on the boundary of the region D at a positive distance d from E and let $f_1(z)$ denote a function that is analytic in D. The function $f_2(z) = f_1(z) + \epsilon/(z - z_0)$, where ϵ is a given positive number, is also analytic in D. On the curve E, these functions differ by $\epsilon/(z - z_0)$, which does not exceed in absolute value the ratio ϵ/d; that is, $|f_2(z) - f_1(z)| \leqslant \epsilon/d$ everywhere on E. The ratio ϵ/d can be made arbitrarily small by choosing ϵ sufficiently great. However, the difference $f_2(z) - f_1(z) = \epsilon/(z - z_0)$ is unbounded in absolute value on the region D as a whole (see also [78, 192, 194, 196]).

Example 6. The inverse gravimetry problem. Suppose that the density of a body is different from that of the surrounding medium. The problem is to determine the shape of the body from the anomalies in the gravitational field strength caused by it on the surface of the earth (see [36–38, 40, 41, 148, 165]).

Let us suppose that the medium under the surface of the earth ($z = 0$) consists of masses with known densities ρ_1 and ρ_2 separated by a boundary $z(x)$ (see Fig. 1).

Suppose that $\tilde{z}(x) = -H$ everywhere except on an interval $a \leqslant x \leqslant b$, on which $\tilde{z}(x) = -H + z(x)$.

*The potential of the gravitational field of the earth satisfies Laplace's equation $\Delta u = 0$.

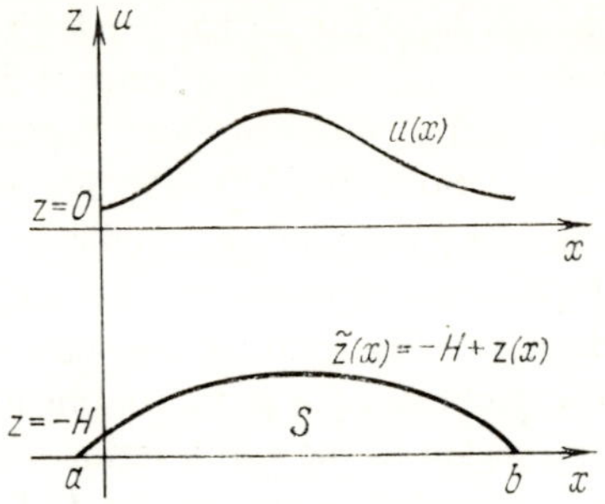

FIG. 1.

Such a configuration of masses causes on the surface of the earth an anomaly in the gravitational field strength

$$\Delta g = -\frac{\partial V}{\partial z}\bigg|_{z=0},$$

where V is the potential due to masses of density $\rho = \rho_2 - \rho_1$ filling the region S shown in Figure 1.

Since

$$V = \int_S \frac{\rho}{2\pi} \ln\left(\frac{1}{r}\right) d\xi\, d\eta, \text{ where } r = \sqrt{(x-\xi)^2 + (z-\eta)^2},$$

we have

$$\Delta g = -\frac{\rho}{2\pi}\int_a^b \int_{-H}^{-H+z(\xi)} -\frac{\partial}{\partial\eta}\ln\frac{1}{r}\, d\xi\, d\eta\,\big|_{z=0} =$$

$$= \frac{\rho}{2\pi}\int_a^b \ln\frac{(x-\xi)^2 + H^2}{(x-\xi)^2 + (H - z(\xi))^2}\, d\xi .$$

An anomaly in the gravitational field strength on the surface of the earth can be measured.

Thus, the problem of determining the function $z(x)$ reduces to solving the nonlinear integral equation of the first kind

$$Az \equiv \int_a^b \ln \frac{(x-\xi)^2 + H^2}{(x-\xi)^2 + (H - z(\xi))^2} \, d\xi = u(x),$$

where $u(x) = (2\pi/\rho)\Delta g$. Here, A is a nonlinear integral operator. It is easy to show that the solution of this equation is unstable under small changes in the right-hand member $u(x)$.

A vast literature [42, 95, 125, 126, 131–140] has been written on questions of uniqueness of the solution of the inverse gravimetry problem.

Let us mention some other classes of mathematical problems that include ill-posed problems: the solution of systems of linear algebraic equations with vanishing determinant (see Chapter III), certain problems in linear programming (see Chapter VIII), problems of minimizing functionals such that convergence of a sequence of values of the functional being minimized to its smallest value does not imply convergence of the minimizing sequence (see Chapter VII), certain optimal control problems (see Chapter VII), and many others [104].

2. An important class of ill-posed problems of great practical value is constituted by problems of planning optimal systems and constructions. We present one of these problems below.

Let us look at a problem of designing physical systems. This class includes problems of designing optical systems with given transmittances (or reflectances). Such systems can be obtained, for example, with the aid of laminated coatings deposited in the form of thin films on the surface of the "backing".

Consider a flat nonhomogeneous plate of infinite extent and thickness H. Let us set up a z-axis perpendicular to the plate with zero point chosen so that, within the plate, $0<z<H$. Let $n(z)$ denote the index of refraction at the point z. Suppose that a monochromatic plane light wave of length $\lambda = 2\pi c/\omega$ is incident normally to the plate. Let us suppose

a) that the half-spaces $\{z<0\}$ and $\{z>H\}$ outside the plate are each homogeneous with indices of refraction n_0 and n_H.

b) that there is no absorption in the plate or in the half-spaces outside it.

Suppose that the electric field of the incident wave in the half-space $\{z<0\}$ has the form $E_{in} = \mathscr{E}_0(z)e^{i\omega t}$, where $\mathscr{E}_0(z) = E_0 e^{-i\frac{\omega}{c}n_0 z}$ is the amplitude of the wave. Under these conditions, the amplitude of the electric field $\mathscr{E}(z)$ satisfies the equation (see [35])

$$\mathscr{E}'' + \frac{\omega^2}{c^2} n_0^2 \mathscr{E} = 0.$$

In this half-space $\{z<0\}$, the total electric field is the resultant of the fields of the incident and reflected waves. The amplitude $\widetilde{\mathscr{E}}_0(z)$ of the total field in this half-space is

$$\widetilde{\mathscr{E}}_0(z) = \mathscr{E}_0(z) + c_1 e^{i\frac{\omega}{c}n_0 z}.$$

The amplitude of the electric field of the wave passing through the plate is $\mathscr{E}_H(z) = c_2 \mathscr{E}_0(z)$.

Inside the plate, the amplitude of the field $\mathscr{E}(z)$ is a solution of the boundary-value problem (see [35])

$$\mathscr{E}'' + \frac{\omega^2}{c^2} n^2(z) \mathscr{E} = 0, \quad 0<z<H, \qquad \text{(I.3.1)}$$

$$\mathscr{E}'(0) - i\frac{\omega}{c} n_0 [\mathscr{E}(0) - 2E_0] = 0, \qquad \text{(I.3.2)}$$

$$\mathscr{E}'(H) + i\frac{\omega}{c} n_0 \mathscr{E}(H) = 0. \qquad \text{(I.3.3)}$$

If $n(z)$ is piecewise-continuous on the interval $(0, H)$, we have, at points of discontinuity z_i,

$$\mathscr{E}(z_i - 0) = \mathscr{E}(z_i + 0), \quad \mathscr{E}'(z_i - 0) = \mathscr{E}'(z_i + 0).$$

The transmission capacity of such a system (the plate) is characterized by the transmittance

$$T = \frac{n(H)}{n_0} \left| \frac{\mathscr{E}(H)}{\mathscr{E}_0(0)} \right|^2. \tag{I.3.4}$$

Obviously, for such a system, this quantity is a function of the length λ of the wave incident to the plate*: $T = T(\lambda)$.

The direct problem consists in finding the transmittance $T(\lambda)$ from given $\mathscr{E}_0$, $n(z)$, and H, and it reduces to solving the boundary-value problem (I.3.1)–(I.3.3) for $\mathscr{E}(z)$.

Frequently, we need to solve the problem inverse to this one. It consists in determining from a given function $\hat{T}(\lambda)$ the system $(n(z), H)$, that is, in finding the index of refraction $n(z)$ of points in the plate and the thickness H of the plate. This problem is called the **synthesis problem.**

It is not well-posed because there may not exist a system $(n(z), H)$ (a plate of thickness H with index of refraction $n(z)$) having the prescribed transmittance $\hat{T}(\lambda)$. It is also possible that more than one system will have that transmittance.

In addition, not every system is realizable in practice.

It is of interest to consider the synthesis problem for laminated systems in which the unknown function $n(z)$ is piecewise-constant. In this case, the index of refraction $n(z)$ has a constant value n_j in the jth layer (for $j = 1, 2, \ldots, N$), the thicknesses of the layers d_j are arbitrary, and the number of layers N is not given. A problem of this kind that is important in practice is that of designing laminated coatings. Let us look in greater detail at the mathematical formulation of this problem, which is the discrete analogue of the problem examined above of designing optical systems.

We shall describe a system consisting of N layers by a $2N$-dimensional vector $x = \{d_1, d_2, \ldots, d_N; n_1, n_2, \ldots, n_N\}$, the components of which are the thicknesses d_j and the indices of

---

*In some cases, one deals not with $T(\lambda)$ but with the reflectance $R(\lambda) = 1 - T(\lambda)$.

refraction n_j of the layers. The solution of problem (I.3.1)–(I.3.3) assigns, by formula (I.3.4), to any $2N$-dimensional vector x a unique transmittance $T(\lambda)$.

Thus, a nonlinear operator

$$A(x, \lambda) = T(\lambda), \qquad \lambda_1 \leqslant \lambda \leqslant \lambda_2 \tag{I.3.5}$$

is defined on some set of $2N$-dimensional vectors x.

Let R^{2N} denote the Euclidean space of $2N$-dimensional vectors and let D_{2N} denote the closed region in it defined by

$$D_{2N} \equiv \{x \in R^{2N};\ d_j \geqslant 0,\ n_{\min} \leqslant n_j \leqslant n_{\max};\ j = 1, 2, \ldots, N\}.$$

We shall estimate the difference between functions $T(\lambda)$ in the metric $L_2(\lambda_1, \lambda_2)$.* Let $\hat{T}(\lambda)$ denote a function in $L_2(\lambda_1, \lambda_2)$ defined on the interval $[\lambda_1, \lambda_2]$.

Let us find

$$\delta_N = \inf_{x \in D_{2N}} \| A(x, \lambda) \| - \hat{T}(\lambda) \|_{L_2}.$$

Obviously, $\delta_1 \geqslant \delta_2 \geqslant \ldots \geqslant \delta_m \geqslant \ldots \geqslant 0$. We shall call the number $\delta_0 = \lim_{m \to \infty} \delta_m$ the limiting attainable accuracy.

The design problem consists in finding an N-layer system with given transmittance $\hat{T}(\lambda)$. Here, we are required to find a system with the *smallest* number of layers N_0 and with *smallest* overall thickness $d = \sum_{j=1}^{N} d_j$. These supplementary requirements are conditioned by the requirement of stability under external disturbances (easily violated when the number of layers is large), by the possibility that the apparatus may be dusty when the coatings are deposited, etc.

The mathematical formulation of this problem consists in finding an approximate solution of the equation

---

*Instead of the L_2-metric, we could take the C-metric.

$$A(x, \lambda) = \hat{T}(\lambda) \qquad (I.3.6)$$

minimizing N and $d = \sum_{j=1}^{N} d_j$ such that $\| A(x, \lambda) - \hat{T}(\lambda) \|_{L_2} \leqslant \delta$, where δ is a given number at least equal to δ_0. This problem can be solved in the following way: By increasing successively by one the number of layers N, we find the smallest value $N = N_0$ for which D_{2N_0} includes a vector x_{N_0} for which $\| A(x_{N_0}, \lambda) - \hat{T}(\lambda) \|_{L_2} \leqslant \delta$.

When we have thus determined the number N_0, we seek, in the class of vectors x in D_{2N_0} satisfying the condition $\| A(x, \lambda) - \hat{T}(\lambda) \|_{L_2} \leqslant \delta$, a vector x minimizing $d = \sum_{j=1}^{N_0} d_j$. A solution of this problem obviously exists since D_{2N_0} is a closed convex set contained in a finite-dimensional Euclidean space. Examples for the numerical solution of such problems are given in [35].

3. The problem of designing optical systems that we have just considered is the problem of determining the coefficients $n(z)$ in equation (I.3.1) from a known functional $T(\lambda)$ of the solution of that equation. It is one of the inverse problems of mathematical physics.

Many other physical problems lead to problems of determining the coefficients in differential equations (ordinary or partial) from certain known functionals of their solutions. An example is the inverse kinematic problem of seismology. Its physical formulation is the following: Let D denote a region bounded by a surface S. A wave process (that is, a process described by the wave equation) generated by sources acting at points of the surface S is considered in D. At points of S different from the wave source points, the time of passage of waves through that region is recorded. We are required to find from these data the speed $a = a(M)$ of propagation of waves within the region as a function of the point M. One can exhibit yet other physical problems leading to inverse problems of this kind. An examination of such problems may be found in [20, 101–103, 111, 141–143].

4. One important problem is the construction of systems of automatic mathematical processing of the results of a physical experiment. One stage in the processing is the solution of inverse problems of the form $Az = u$ for z.

Many modern experimental devices for investigating physical phenomena and objects of various kinds are complicated and expensive complexes (accelerators of elementary particles, devices for obtaining and investigating high-temperature plasma, devices for investigating the properties of matter at very low temperatures, etc.).

The desire for reliable information regarding the phenomenon under investigation or the study of "rare" and "weak" effects frequently necessitates many repetitions of a single experiment. Automation of the performance of an experiment and of registration of its results enables us to obtain in a short time a large volume of necessary information (tens and hundreds of thousands of photographs, oscillograms, readings of detectors, etc.). To obtain from this information the necessary characteristics of the phenomenon or object being studied, we subsequently have to process the results of its observations. In many cases, this processing must be done almost simultaneously with the performance of the experiment or with only a small permissible lag. In such a case, the processing necessitates going over a large amount of information and can be done only with a computer.

For a large class of experimental set-ups, we can single out the following steps in the processing of observational results [66].

First step. Information is taken from the recording device or from a constant carrier (for example, a photographic cell), converted to a numerical code, and entered in the memory of the computer.

Second step. The results of observation are statistically processed with an estimate of the degree of reliability. Normalizing measurements (background, calibrational, dosimetric, etc.) are allowed for.

Third step. The results obtained at the second step of the processing are interpreted. Usually, this interpretation consists in estimating the sought characteristics of the model of the phenomenon or object being studied.

What is usually recorded in a physical experiment is not the parameters z of interest to us but certain of their manifestations $u = Az$. Therefore, the interpretation problem usually reduces to solving an equation $Az = u$. In many cases, this problem is ill-posed.

A program for carrying out the entire processing complex (from the first through the third step) is usually called a **complete system of mathematical processing** of observational results or, more briefly, a **processing system**. A processing system can operate both in a "dialogue regime" and automatically, without the intervention of a human being in the intermediate stages. This procedure for automatic mathematical processing of observational results requires the use, in the processing system, of algorithms for solving the equation $Az = u$ (including ill-posed problems) that are easily realized on a computer.

A description is given in [33, 176] of a system of automatic mathematical processing of the results of a physical experiment on the interaction of γ-quanta with neutrons and protons.

5. *Inverse* problems constitute a broad class of ill-posed problems arising in physics, technology, and other branches of learning.

Suppose that the object or phenomenon in question is characterized by an element z_T (a function or vector) belonging to a set F. Frequently, z_T does not lend itself to direct study and some manifestation of it $Az_T = u_T$ is investigated (here, $u_T \in AF$, where AF is the image of the set F under the mapping executed by the operator A). Obviously, the equation

$$Az = u \tag{I.3.7}$$

has a solution on F only for those elements u that belong to the set AF. The element u_T is usually obtained from measurements and hence we know it only approximately. Let $\tilde{u}$ denote its approximate value. In such cases, we can speak only of finding an approximate solution (that is, close to z_T) of the equation

$$Az = \tilde{u}. \tag{I.3.8}$$

Here, $\tilde{u}$ generally does not belong to the set AF. Frequently, the operator A is such that the inverse operator A^{-1} is not continuous (for example, when A is a completely continuous operator, in particular, the integral operator of Example 1). Under these conditions, we cannot take for the approximate solution the exact solution of equation (I.3.7) with approximate right-hand member; that is, we cannot take for the approximate solution the element $z = A^{-1}\tilde{u}$ for two reasons: First, such a solution may not exist on the set F since $\tilde{u}$ may not belong to the set AF, so that requirement 1) of a well-posed problem is not satisfied. Second, even if such a solution does exist, it will not possess the property of stability since the inverse operator A^{-1} is not continuous whereas stability of a solution of the problem (I.3.7) is usually a consequence of its physical determinacy and hence an approximate solution must possess this property. Thus, requirement 3) of a well-posed problem is not satisfied. Consequently, the problem (I.3.7) is ill-posed.

In many cases, the absence of stability makes difficult a physical interpretation of the measurement results. Satisfying this condition is also necessary for the use of numerical methods of solution on the basis of approximate initial data. Thus, in the case of inverse problems we are confronted with the basic question: *what do we mean by "approximate solution" of such problems*? If an answer is given to this question, the problem then arises of finding algorithms for constructing approximate solutions that will possess the property of stability with respect to small changes in the initial data.

6. In what follows, our basic attention will be given to methods of solving ill-posed problems of the form (I.3.7). These include systems of linear algebraic equations and Fredholm integral equations of the first kind. In the first case, the operator A is a matrix whose elements are the coefficients in the system; in the second, the operator A is the integral operator

$$Az \equiv \int_a^b K(x, s)\, z(s)\, ds.$$

7. Suppose that the elements z and u in equation (I.3.7) are functions (scalar or vector) of the point M and the time t and that the operator A is a linear operator determined by the nature of the transformation from z to u. If we know the reaction (response) of the transformer to the Dirac delta function, which becomes infinite when $M = P$ and $t = \tau$, that is, the function $A\delta(M, P; t, \tau) = K(M, P; t, \tau)$, then, for an arbitrary function $z(M, t)$ in F, we have

$$Az \equiv \int K(M, P; t, \tau) z(P, \tau)\, dv_P d\tau,$$

where the integral is over the entire domain of definition of the function $z(P, t)$. The function $K(M, P; t, \tau)$ is called the **impulse transfer function** or, more briefly, the **impulse function** of the transformer (system). It is also often called the **apparatus function** of the system. In this case, equation (I.3.7) reduces to an integral equation of the first kind.

Let us consider, for example, the problem of studying the spectral composition of a beam of light (the spectroscopy problem). Suppose that the observed radiation is nonhomogeneous and that the distribution of the energy density over the spectrum is characterized by a function $z(s)$, where s is the frequency (or energy). If we pass the beam through a measuring apparatus, we obtain an experimental spectrum $u(x)$. Here, x may be the frequency and it may also be expressed in terms of voltage or current of the measuring device. If the measuring device is linear, then the functional relationship between $z(s)$ and $u(x)$ is given by

$$Az \equiv \int_a^b K(x, s) z(s)\, ds = u(x),$$

where $K(x, s)$ is the apparatus function, assumed to be known. It represents the experimental spectrum (as a function of x) when a monochromatic beam of frequency s and unit intensity (this is the delta function $\delta(s-x)$) hits the device. The numbers a and b are the ends of the spectrum.

8. If z depends only on the time and if the transformer (apparatus) is homogeneous with respect to time, then the function $K(M, P; t, \tau)$ depends only on the difference $t-\tau$:

$$K(M, P; t, \tau) = K(t-\tau).$$

In this case, the operator Az has the form

$$Az \equiv \int_{-\infty}^{\infty} K(t-\tau) z(\tau) d\tau,$$

and equation (I.3.7) becomes

$$\int_{-\infty}^{\infty} K(t-\tau) z(\tau) d\tau = u(t). \qquad \text{(I.3.9)}$$

This is an integral equation of the first kind of the convolution type. Examples of problems leading to equation (I.3.9) are the following:

a) the problem of determining the form of a radio impulse $z(t)$ emitted by a source from the results of its reception at great distances from the source $u(t)$ when we know the impulse function of the trace of the distribution $K(t)$. The equation for $z(t)$ has the form

$$Az \equiv \int_{0}^{t} K(t-\tau) z(\tau) d\tau = u(t).$$

b) the problem of determining the form of an electrical impulse at the input of a cable $z(t)$ from measurements of it at the output of the cable $u(t)$:

$$\int_0^t K(t-\tau)\, z(\tau)\, d\tau = u(t),$$

where $K(t)$ is the impulse function of the cable (see Chapter II).

c) Calculation of the derivatives of a function that is known only approximately. For the nth derivative $z(x)$ of a function $u(x)$ we have the equation

$$\int_{x_0}^{x} \frac{1}{(n-1)!} (x-t)^{n-1} z(t)\, dt = u(x).$$

d) Automatic control problems, for example, the determination of the transfer functions $k(t)$ of linear transformers from the input and output signals $x(t)$ and $y(t)$:

$$Ak \equiv \int_0^t x(t-\tau)\, k(\tau)\, d\tau = y(t).$$

There are also many others (see [148, 164]).

CHAPTER I

THE SELECTION METHOD. QUASISOLUTIONS

The possibility of determining approximate solutions of ill-posed problems that are stable under small changes in the initial data is based on the use of supplementary information regarding the solution. Various kinds of supplementary information are possible. In the first category of cases, the supplementary information, which is of a quantitative nature, enables us to narrow the class of possible solutions, for example, to a compact set, and the problem becomes stable under small changes in the initial data. In the second category of cases, to find approximate solutions that are stable under small changes in the initial data, we use only qualitative information regarding the solution (for example, information regarding its smoothness (see [156–161, 175]).

In the present chapter, we shall look at the selection method, which has broad practical application, the quasisolution method, the method of replacement of the original equation with an equation close to it, and the quasiinversion method. As an ill-posed problem, we shall consider the problem of solving the equation

$$Az = u \tag{1.0.1}$$

for z, where u belongs to a metric space U and z belongs to a metric space F. The operator A maps F onto U. It is assumed that A has an inverse operator A^{-1} though the latter is not in general continuous. When the operator A possesses these properties, we shall call equation (1.0.1) an **operator equation of the first kind** or, more briefly, an **equation of the first kind**.

§1. The selection method of solving ill-posed problems.

1. A method of solving equation (1.0.1) approximately that is widely used in computational work is the selection method. It consists in calculating the operator Az for elements z belonging to some given subclass $M \subset F$ of possible solutions; that is, we solve the direct problem. As an approximate solution, we take an element z_0 belonging to the set M for which the difference $\rho_U(Az, u)$ attains its minimum:

$$\rho_U(Az_0, u) = \inf_{z \in M} \rho_U(Az, u).$$

Suppose that we know the right-hand member of equation (1.0.1) exactly: $u = u_T$, and that we are required to find its solution z_T. Usually, we take for M a set of elements z depending on a finite number of parameters varying within finite limits in such a way that M is a closed set contained in a finite-dimensional space. If the desired exact solution z_T of equation (1.0.1) belongs to the set M, then $\inf_{z \in M} \rho_U(Az, u) = 0$ and this infimum is attained with the exact solution z_T. If equation (1.0.1) has a unique solution, the element z_0 minimizing $\rho_U(Az, u)$ is uniquely defined.

In practice, minimization of $\rho_U(Az, u)$ is done only approximately, and our first fundamental question about the selection method is as follows: If $\{z_n\}$ is a sequence of elements of M such that $\rho_U(Az_n, u) \to 0$ as $n \to \infty$, can we assert that $\rho_F(z_n, z_T) \to 0$ as $n \to \infty$, that is, that $\{z_n\}$ converges to z_T?

2. The attempt to find a basis for the success of the selection method has led to some general functional requirements restricting

the class M of possible solutions for which the selection method is stable and $z_n \to z_T$ [155]. These requirements consist in compactness of the set M and are based on the following topological lemma:

Lemma. *Suppose that a compact (in itself) subset F of a metric space F_0 is mapped onto a subset U of a metric space U_0. If the mapping $F \to U$ is continuous and one-to-one, the inverse mapping $U \to F$ is also continuous.*

Proof. Let φ denote the function that maps any element z of F into an element u of U and let ψ denote the inverse mapping from U onto F.

Let us take an arbitrary element u_0 of U. Let us show that the function $\psi(u)$ is continuous at u_0. Suppose that this is not the case. Then, there exists a positive number ϵ_1 such that, for every position number δ, there exists an element $\tilde{u}$ of U such that $\rho_U(\tilde{u}, u_0) < \delta$ though $\rho_F(\tilde{z}, z_0) \geqslant \epsilon_1$. Here, $\tilde{z} = \psi(\tilde{u})$ and $z_0 = \psi(u_0)$.

Let us take a sequence $\{\delta_n\}$ of values of δ that converges to 0 as $n \to \infty$. For every δ_n, there exists an element $\tilde{u}_n$ such that

$$\rho_U(\tilde{u}_n, u_0) < \delta_n \text{ and } \rho_F(\tilde{z}_n, z_0) \geqslant \epsilon_1, \text{ where } \tilde{z}_n = \psi(\tilde{u}_n).$$

Obviously, the sequence $\{\tilde{u}_n\}$ converges to the element u_0. Since all the elements $\tilde{z}_n$ belong to the compact set F, the sequence $\{\tilde{z}_n\}$ has a subsequence that converges to an element $\tilde{z}_0$ of F:

$$\{\tilde{z}_{n_k}\} \to \tilde{z}_0.$$

Here, $\tilde{z}_0 \neq z_0$ since $\rho_F(\tilde{z}_{n_k}, z_0) \geqslant \epsilon_1$. Corresponding to this subsequence is a sequence of elements

$$\tilde{u}_{n_k} = \varphi(\tilde{z}_{n_k})$$

of U that converges (in the sense of continuity of the mapping φ) to the element $\tilde{u}_0 = \varphi(\tilde{z}_0)$ and is a subsequence of the sequence $\{\tilde{u}_n\}$. Since this last sequence converges to $u_0 = \varphi(z_0)$, we have

$$\tilde{u}_0 = \varphi(\tilde{z}_0) = u_0 = \varphi(z_0).$$

Since the mapping $F \to U$ is one-to-one, it follows that $\tilde{z}_0 = z_0$. Since we showed above that $\tilde{z}_0 \neq z_0$, we have reached a contradiction. This completes the proof of the lemma.

Thus, the minimizing sequence $\{z_n\}$ in the selection method converges to z_T as $n \to \infty$ if z_T belongs to a compact class M of possible solutions.

Suppose that, instead of the exact right-hand member u_T, we have an element u_δ such that $\rho_U(u_\delta, u_T) \leqslant \delta$. If u_δ belongs to the set AM (the image of the set M under the operator A) and M is compact, we can use the selection method to find an approximate solution z_n^δ of the equation $Az = u_\delta$. It will also be an approximation of the solution z_T of the equation $Az = u_T$ since the inverse operator A^{-1} is continuous on AM. In finding z_n^δ as an approximation of z_T, we need to keep in mind the level of error in the right-hand member since

$$\rho_U(Az_n^\delta, u_T) \leqslant \rho_U(Az_n^\delta, u_\delta) + \rho_U(u_\delta, u_T).$$

3. On the basis of what has been said, M. M. Lavrent'yev formulated [99] the concept of a **well-posed problem in the sense of Tikhonov**. As applied to equation (1.0.1), a problem is said to be well-posed in the sense of Tikhonov if we know that, corresponding to the exact solution $u = u_T$, there exists a unique solution z_T of equation (1.0.1) (so that $Az_T = u_T$) belonging to a given compact set M. In this case, the operator A^{-1} is continuous on the set $N = AM$ and, if we know not the element u_T but an element u_δ such that $\rho_U(u_T, u_\delta) \leqslant \delta$ and $u_\delta \in N$, then for an approximate solution of equation (1.0.1) with right-hand member $u = u_\delta$ we can take the element $z_\delta = A^{-1}u_\delta$. Since $u_\delta \in N$, this z_δ approaches z_T as $\delta \to 0$. A set F_1 (contained in F) on which the problem of finding a solution of equation (1.0.1) is well posed is called a **well-posedness class**. Thus, if the operator A is continuous and one-to-one, the compact set M to which z_T is restricted is a well-posedness class for equation (1.0.1). Hence, if the problem (1.0.1) is well-posed in the sense of Tikhonov and the right-hand member u belongs to AM, then the selection method

check with Burgraff

can be used successfully to solve the problem. Thus, we have given a definitive answer to the first question.

Let us look at the problem of solving the Fredholm integral equation of the first kind

$$A_1 z \equiv \int_a^b K(x, s) z(s)\, ds = u(x), \;\; u(x) \in L_2 \qquad (1.1.1)$$

on a set M_1 of decreasing (or increasing) uniformly bounded functions; that is, for some B, we have $|z(s)| \leqslant B$. This problem is well-posed in the sence of Tikhonov since the set M_1 is compact in the space L_2 [43].

To see this, let us take an arbitrary sequence $E = \{z_1(s), z_2(s), \ldots, z_n(s), \ldots\}$ of functions in M_1. According to Helly's choice theorem,* there exists a subseqence

$$E_1 \equiv \{z_{n_1}(s), z_{n_2}(s), \ldots, z_{n_k}(s), \ldots\}$$

of the sequence E and a function $\bar{z}(s)$ in $M_1 \subset L_2$ such that

$$\lim_{n_k \to \infty} z_{n_k}(s) = \bar{z}(s)$$

except possibly at a denumerable set of points of discontinuity of the function $\bar{z}(s)$. As we know,** pointwise convergence of the subsequence E_1 to the function $\bar{z}(s)$ everywhere except possibly at a denumerable set of points implies convergence of the subsequence E_1 to the function $\bar{z}(s)$ in the metric of L_2.

Thus, as an approximate solution of equation (1.1.1) on the set M_1 with only approximately known right-hand member $\tilde{u} \in AM_1$,

---

*E. C. Titchmarch, *Eigenfunction Expansions Associated with Second-Order Differential Equations*, Oxford University Press, part 2, 1958.

**M. A. Krasnosel'skiy, P. P. Zabreyko, Ye. I. Pustyl'nik, and P. Ye. Sobolevskiy, *Integral'nyye operatory v prostranstvakh summiruyemykh funktsiy* (Integral operators in spaces of summable functions), Moscow, Nauka, 1966.

we can take the exact solution of that equation with right-hand member $u = \tilde{u}$. As we know, this last problem is equivalent to the problem of finding on the set M_1 the function minimizing the functional

$$N[z, \tilde{u}] = \| A_1 z - \tilde{u} \|^2_{L_2}.$$

Suppose that $\rho_U(u_T, \tilde{u}) \leqslant \delta$. Then, as approximate solution of equation (1.1.1), we can take the function z_δ for which

$$\| A_1 z_\delta - \tilde{u} \|^2_{L_2} \leqslant \delta^2. \qquad (1.1.2)$$

If we replace the integral operator A_1 with an operator representing summation over a fixed grid with n nodes and if we denote by z_i the values of the unknown function at the nodes, the problem of constructing an approximate solution of equation (1.1.1) reduces to the problem of finding a finite-dimensional vector minimizing the functional $N[z, \tilde{u}]$ and satisfying equation (1.1.2). This is a linear programming problem.

In certain cases, we can actually exhibit compact well-posedness classes. This makes it possible to construct stable approximate solutions.

4. Because of error in the initial data, the element u may fail to belong to the set AM. Under these conditions, equation (1.0.1) does not have a (classical) solution and the question arises as to what we mean by an approximate solution of equation (1.0.1).

In this case, we introduce the concept of a quasisolution. If the set M is compact, the selection method enables us to find an approximation to the quasisolution.

In the following section, we shall treat quasisolutions in detail.

§2. Quasisolutions.

1. Suppose that the operator A in equation (1.0.1) is completely continuous. In those cases, mentioned in §1, in which we are seeking a solution on a compact subset M of F and the right-handed member u of the equation belongs to the set $N = AM$, it is possible to use the formula

$$z = A^{-1}u \qquad (1.2.1)$$

to construct a solution of equation (1.0.1) that is stable with respect to small changes in the right-hand member u of the approximate equation.

Most of the time, we do not have an effective criterion for determining whether the element u belongs to the set N or not. This needs to be assumed known in advance. In practical problems, we often know not the exact value of the right-hand member u_T but an approximate value of it $\tilde{u}$, which may not belong to the set $N = AM$. In these cases, we cannot use formula (1.2.1) to construct an approximate solution of equation (1.0.1) since the symbol $A^{-1}\tilde{u}$ may be meaningless.

2. The attempt to avoid the difficulties associated with the absence of a solution of equation (1.0.1) in the case of an inexact right-hand member u led V. K. Ivanov [71, 72] to the concept of a quasisolution of equation (1.0.1), which is a generalization of the concept of a solution of that equation.

An element of $\tilde{z} \in M$ minimizing, for given u, the functional $\rho_U(Az, u)$ on the set M is called a **quasisolution** of equation (1.0.1) on M:

$$\rho_U(A\tilde{z}, u) = \inf_{z \in M} \rho_U(Az, u).$$

If M is a compact set, a quasisolution obviously exists for every $u \in U$. If in addition $u \in AM$, then the quasisolution $\tilde{z}$ coincides with the usual (exact) solution of equation (1.0.1). There may be more than one quasisolution. In such a case, by a quasisolution we shall mean any element of the set D of quasisolutions.

It is possible to exhibit sufficient conditions for a quasisolution to be unique and to depend continuously on the right-hand member u.

Let us recall the definition. Let y denote an element and Q a subset of the space U. An element q of the set Q is called a **projection** of the element y onto the set Q (and we write $q = Py$) if

$$\rho_U(y, q) = \rho_U(y, Q),$$

where

$$\rho_U(y, Q) = \inf_{h \in Q} \rho_U(y, h).$$

Theorem 1. *If the equation $Az = u$ can have more than one solution on a compact set M and if the projection of each element u of U onto the set $N = AM$ is unique, then the quasisolution of equation* (2.0.1) *is unique and depends continuously on the right-hand member u.*

Proof. Suppose that $\tilde{z}$ is a quasisolution and that $\tilde{u} = A\tilde{z}$. Obviously, $\tilde{u}$ is a projection of the element u onto the set $N = AM$. By the hypothesis of the theorem, it is unique. This fact and the fact that the mapping of the set M onto the set N is one-to-one imply uniqueness of the quasisolution $\tilde{z}$.

Obviously, $\tilde{z} = A^{-1}\tilde{u} = A^{-1}Pu$. By virtue of the lemma on the continuity of the inverse mapping of a compact set (see §1), the operator A^{-1}) is continuous on N. The projection operator P is continuous* on U. Therefore, $A^{-1}P$ is a continuous operator on U and hence the quasisolution $\tilde{z}$ depends continuously on the right-hand member u.

Thus, all the conditions for well-posedness are restored when we shift to the quasisolution; that is, the problem of finding the quasisolution of equation (1.0.1) on the compact set M is well-posed.

If the solution of equation (1.0.1) is not unique, then the quasisolutions constitute a set D of elements of the compact set M. In this case, without the restrictions listed in Theorem 1 on the set N, we have continuous dependence of the set of quasisolutions D on u in the sense of continuity of multiple-valued mappings. It is not difficult to prove this assertion (see [72, 106]) though its proof would require the introduction of a number of new

---

*L. A. Lyusternik and V. I. Sobolev, *Elements of Functional Analysis,* Ungar, New York, 1961 (translation of *Elementy funktsional'nogo analiza*).

concepts and we shall not stop for this. For the case in which equation (1.0.1) is linear, we can easily obtain more general results contained in the following theorem [72]:

Theorem 2. *Suppose that equation* (1.0.1) *is linear, that the homogeneous equation $Az = 0$ has only the zero solution, that the set M is convex, and that every sphere in the space U is strictly convex. Then, a quasisolution of equation* (1.0.1) *on the compact set M is unique and it depends continuously on the right-hand member u.*

Proof. Let $\tilde{z}$ denote a quasisolution and suppose that $\tilde{u} = A\tilde{z}$. Since the set M is convex, the set $N = AM$ is also convex by virtue of the linearity of the operator A. Obviously, $\tilde{u}$ is the projection of the element u onto the set N. Since every sphere in the space U is assumed to be strictly convex, the projection $\tilde{u}$ is unique. The remainder of the proof is the same as for Theorem 1.

3. Suppose that F and U are Hilbert spaces, that $M \equiv S_R$ is the ball $(\|z\| \leqslant R)$ in the space F, and that A is a completely continuous operator.

In this case, a quasisolution of equation (1.0.1) can be represented in the form of a series of eigenfunctions or eigenvectors φ_n of the operator A^*A, where the asterisk denotes the adjoint operator.

We know that A^*A is a self-adjoint positive completely continuous operator from F into F. Let $\lambda_1 \geqslant \lambda_2 \geqslant \ldots \lambda_n \geqslant \ldots$ denote the complete system of its eigenvalues and let $\varphi_1, \varphi_2, \ldots, \varphi_n, \ldots$ denote the corresponding complete orthonormalized system of its eigenfunctions or eigenvectors. The element A^*u can be represented by a series of the form

$$A^*u = \sum_{n=1}^{\infty} b_n \varphi_n. \qquad (1.2.2)$$

Under these conditions, we have [72]:

Theorem 3. *The quasisolution of equation (1.0.1) on the set S_R is expressed by*

$$\tilde{z} = \sum_{n=1}^{\infty} \frac{b_n}{\lambda_n} \varphi_n \tag{1.2.3}$$

if

$$\sum_{n=1}^{\infty} \frac{b_n^2}{\lambda_n^2} < R^2 \tag{1.2.4}$$

but by

$$\tilde{z} = \sum_{n=1}^{\infty} \frac{b_n}{\beta + \lambda_n} \varphi_n$$

if

$$\sum_{n=1}^{\infty} \frac{b_n^2}{\lambda_n^2} \geqslant R^2. \tag{1.2.5}$$

Here, β is a root of the equation

$$\sum_{n=1}^{\infty} \frac{b_n^2}{(\lambda_n + \beta)^2} = R^2. \tag{1.2.6}$$

Proof. The quasisolution minimizes the functional

$$\rho_U^2 (Az, u) = (Az - u, Az - u), \tag{1.2.7}$$

for which the Euler equation has the form*

$$A^* Az = A^* u. \tag{1.2.8}$$

---

*V. I. Smirnov, *Course of Higher Mathematics*, Vol. V, Addison-Wesley, Reading, Mass., 1964 (translation of *Kurs vysshey matematiki*).

We seek the solution of this equation in the form of a series of the φ_n:

$$z = \sum_{n=1}^{\infty} c_n \varphi_n. \tag{1.2.9}$$

Substituting this series into equation (1.2.8) and using the expansion (1.2.2), we find $c_n = b_n/\lambda_n$. Consequently, inequality (1.2.4) means that $\|z\| < R$ and it is a question of finding the unconditional extremum of the functional (1.2.7). The series (1.2.3) is then the solution of the problem.

On the other hand, if inequality (1.2.5) holds, this means that $\|z\| \geqslant R$ and we need to solve the problem for a conditional extremum of the functional (1.2.7) under the condition that $\|z\|^2 = R^2$. By using the method of undetermined Lagrangian multipliers, we can reduce the problem to that of finding the unconditional extremum of the functional

$$(Az - u, Az - u) + \alpha (z, z).$$

and we can then reduce this problem to that of finding the corresponding solution of the Euler equation $A^*Az + \alpha z = A^*u$. Substituting for z the expression given by (1.2.9) and using the expansion (1.2.2), we find

$$c_n = \frac{b_n}{\alpha + \lambda_n}.$$

We determine the parameter α from the condition $\|z\|^2 = R^2$, which is equivalent to (1.2.6).

§3. Approximate determination of quasisolutions.

In the preceding section, we saw that finding a quasisolution involves finding an element in an infinite-dimensional space. For approximate determination of a quasisolution, it is natural to shift to a finite-dimensional space. There is a fairly general approach to approximate determination of quasisolutions of equations (1.0.1) [60, 72], in which A is a completely continuous operator.

We shall assume that the sufficient conditions listed in section 2 for existence of a unique quasisolution on a given set M are satisfied; that is, we shall assume that the set M is convex and compact and that a sphere in the space U is strictly convex.

Let $M_1, M_2, \ldots$ denote a sequence of closed compact sets M_n such that

$$M_1 \subset M_2 \subset \ldots \subset M_n \subset \ldots$$

and the closure of their union $\bigcup_{n=1}^{\infty} M_n$ coincides with M. A quasisolution of equation (1.0.1) exists on each set M_n, but it may not be unique. We denote by T_n the set of all quasisolutions on the set M_n.

Let us show that we can take for an approximation of the quasisolution $\tilde{z}$ on the set M any element $\tilde{z}_n$ in T_n. Here,

$$\lim_{n\to\infty} \rho_F(\tilde{z}_n, \tilde{z}) = 0.$$

Define $N_n = AM_n$ and let B_n denote the set of projections of the element u onto the set N_n. Obviously, $B_n = AT_n$ and $N_1 \subseteq N_2 \subseteq \ldots \subseteq N_n$. Then,

$$\rho_U(u, N_1) \geqslant \ldots \geqslant \rho_U(u, N_n) \geqslant \ldots \geqslant \rho_U(u, N) = \\ = \rho_U(u, A\tilde{z}). \quad (1.3.1)$$

Since the set $\bigcup_{n=1}^{\infty} N_n$ is everywhere dense in N, for every positive ϵ there exists a number $n_0(\epsilon)$ such that, for every $n > n_0(\epsilon)$,

$$\rho_U(u, N_n) < \rho_U(u, N) + \epsilon. \quad (1.3.2)$$

It follows from (1.3.1) and (1.3.2) that

$$\lim_{n\to\infty} \rho_U(u, N_n) = \rho_U(u, N). \qquad (1.3.3)$$

Since

$$\rho_U(u, N_n) = \rho_U(u, B_n),$$

we have

$$\lim_{n\to\infty} \rho_U(u, B_n) = \rho_U(u, A\tilde{z}). \qquad (1.3.4)$$

Each set B_n is compact since it is a closed subset of the compact set N_n. Therefore, there exists in B_n an element y_n such that

$$\rho_U(y_n, u) = \inf_{y\in B_n} \rho_U(y, u).$$

The sequence $\{y_n\}$ has at least one cluster point belonging to N since N is a compact set. Let y_0 denote a cluster point of the set $\{y_n\}$ and let $\{y_{n_k}\}$ denote a subsequence that converges to y_0; that is,

$$\lim_{n_k\to\infty} \rho_U(y_{n_k}, y_0) = 0.$$

It follows from (1.3.3) and (1.3.4) that

$$\rho_U(u, y_0) = \lim_{n_k\to\infty} \rho_U(u, y_{n_k}) =$$
$$= \lim_{n_k\to\infty} \rho_U(u, B_{n_k}) = \rho_U(u, A\tilde{z}) = \rho_U(u, N).$$

Thus,

$$\rho_U(u, y_0) = \rho_U(u, N).$$

This fact and the uniqueness of the quasisolution on the set M imply that

$$y_0 = A\tilde{z}.$$

Since y_0 is an arbitrary cluster point of the set $\{y_n\}$, the sequence $\{y_n\}$ converges to $A\tilde{z}$. This means that we can take as approximation of the quasisolution any element $\tilde{z}_n$ in the set T_n since, by virtue of the lemma of §1, $\tilde{z}_n \to \bar{z}$ as $n \to \infty$.

If we take for M_n an n-dimensional set, the problem of finding an approximate quasisolution on the compact set M reduces to minimizing the functional $\rho_U(Az, u)$ on the set M_n, that is, to finding the minimum of a function of n variables.

Quasisolutions have also been studied in [52, 53, 58, 59, 61, 74, 106, 109].

§4. Replacement of the equation $Az = u$ with an equation close to it.

Equations of the form (1.0.1) in which the right-hand member u does not belong to the set $N = AM$ have been studied by M. M. Lavrent'yev [97–99]. To him belongs the idea of replacing the original equation (1.0.1) with an equation that in some sense is close to it and for which the problem of finding the solution is stable under small changes in the right-hand member and solvable for an aribtrary right-hand member u belonging to U. In the simplest case, this is done in the following way:

Suppose that $F = U = H$ are Hilbert spaces, that A is a bounded, positive, self-adjoint linear operator, that $S_R \equiv \{x, \|x\| \leqslant R, x \in F\}$ is the ball of radius R in the space F, and that B is a completely continuous operator defined, for every $R > 0$, on S_R. As well-posedness class M, we take the set $D_R = BS_R$, that is, the image of the sphere S_R under the operator B. It is assumed that the sought exact solution z_T of equation (1.0.1) with right-hand member $u = u_T$ exists and belongs to the set D_R. Equation (1.0.1) is replaced with the equation

$$(A + \alpha E) z \equiv Az + \alpha z = u, \tag{1.4.1}$$

where α is a positive numerical parameter. With appropriate choice of the parameter α, the solution of equation (1.4.1)

$$z_\alpha = (A + \alpha E)^{-1} u, \tag{1.4.2}$$

is taken for the approximate solution of equation (1.0.1). Here, E is the identity operator.

Remark. To estimate the deviation $\rho_F(z_T, z_\delta)$ of the approximate solution from the exact one, we can use the modulus of continuity ω of the inverse operator on N.

Suppose that u_1 and u_2 belong to N and that $\rho_U(u_1, u_2) \leqslant \delta$. Then,

$$\omega(\delta, N) = \sup_{u_1, u_2 \in N} \rho_F(A^{-1}u_1, A^{-1}u_2).$$

Obviously, if $\rho_U(u_T, u_\delta) \leqslant \delta$ and $z_\delta = A^{-1} u_\delta$, then

$$\rho_F(z_T, z_\delta) \leqslant \omega(\delta, N).$$

Let us return to equation (1.4.1). If $\|Az\| \leqslant \delta$ and $\omega(\delta, D_R) = \sup_{D_R} \|z\|$, we can easily obtain an estimate of the deviation of z_α from z_T. Obviously,

$$\|z_\alpha - z_T\| \leqslant \|\bar{z}_\alpha - z_T\| + \|z_\alpha - \bar{z}_\alpha\|, \tag{1.4.3}$$

where

$$\bar{z}_\alpha = (A + \alpha E)^{-1} u_T.$$

Consequently,

$$\|z_\alpha - z_T\| \leqslant \omega(\delta, D_R) + \frac{\delta}{\alpha}. \tag{1.4.4}$$

If we know the modulus of continuity $\omega(\delta, D_R)$ or a majorant of it, we can use (1.4.4) to find the value of the parameter α as a function of δ that will minimize the right-hand member of inequality (1.4.4).

§5. The method of quasiinversion.

1. We know that the Cauchy problem for the heat flow equation with negative values of the time is unstable under small changes in the initial values. The instability remains in cases in which the solution is subject to certain extra boundary conditions. The method of quasiinversion [104] has been developed for a stable solution of such problems. Let us describe the essential features of that method for a very simple heat-flow equation without justifying our steps. A detailed treatment, applicable to a broader class of problems, is contained in [104].

2. Let us look at the direct problem. Let D denote a finite region in n-dimensional Euclidean space R^n of points $x = (x_1, x_2, \ldots, x_n)$. Suppose that D is bounded by a piecewise-smooth surface S. Let $\varphi(x)$ denote a given continuous function defined on D. With t denoting the time, the direct problem consists in finding a solution $u = u(x, t)$, in the region $G = \{x \in D, t > 0\}$, of the equation

$$\frac{\partial u}{\partial t} - \Delta u = 0 \tag{1.5.1}$$

that satisfies the boundary condition

$$u(x, t) = 0 \quad \text{for} \quad x \in S \tag{1.5.2}$$

and the initial condition

$$u(x, 0) = \varphi(x). \tag{1.5.3}$$

Here,

$$\Delta u = \sum_{k=1}^{n} \frac{\partial^2 u}{\partial x_k^2}.$$

We know that this problem has a solution. To every function

$\varphi(x) \in C$ there corresponds a solution of the problem (1.5.1)–(1.5.3). Let us denote it by $u(x, t; \varphi)$.

The inverse problem consists in finding the function $\varphi(x)$ when we know the function $u(x, t; \varphi)$. In actual problems, the function $u(x, t; \varphi)$ is usually found empirically and hence known only approximately. Let us assume that $u \in L_2$. Such a function may fail to correspond to any "initial" function $\varphi(x)$. Thus, there may not be a solution of the inverse problem in the class of functions C. Therefore, let us look at the problem of finding a generalized solution of the inverse problem.

Suppose that we are given a positive number T and a function $\psi(x)$ belonging to L_2 and defined on D. The functional

$$f(\varphi) = \int_D | u(x, T; \varphi) - \psi(x) |^2 \, dx$$

is defined on the set of functions $\varphi(x)$ in the class C. We shall refer to a function $\varphi(x)$ for which

$$f_0 = \inf_{\varphi \in C} f(\varphi)$$

as a **generalized solution** of the inverse problem.

Remark. A "natural" approach to solution of this problem is to choose the function $\varphi(x)$ in such a way that $f(\varphi) = 0$. To do this, it is sufficient to find the solution of the direct problem

$$\frac{\partial u}{\partial t} - \Delta u = 0;$$
$$u(x, t) = 0 \quad \text{for } x \in S, \ 0 < t < T;$$
$$u(x, T) = \psi(x)$$

and then set $\varphi(x) = u(x, 0)$. But this problem is not in general solvable for a given function $\psi(x)$ in L_2. Furthermore, it is unstable under small changes in the function $\psi(x)$.

On a certain class of generalized functions $\varphi(x)$, we have $f_0 = 0$ (see [104]). Therefore, we shall examine the problem of finding an approximate value of f_0 with a given error level:

For given $\epsilon > 0$, find a function $\varphi_\epsilon(x)$ for which $f(\varphi_\epsilon) \leqslant \epsilon$.

This problem is solved by the method of quasiinversion.

The idea of the method of quasiinversion consists in finding an operator B_α close to the heat-flow operator $\partial/\partial t - \Delta$ such that, when it is substituted for the heat-flow operator, the problem is stable for $t < T$:

$$B_\alpha u_\alpha = 0, \quad x \in D, \quad t < T, \quad \alpha > 0;$$
$$u_\alpha(x, T) = \psi(x);$$
$$u_\alpha(x, t) = 0 \text{ for } x \in S, \ t < T$$

When we solve this problem, we get

$$\varphi(x) = u_\alpha(x, 0).$$

Usually, for the operator B_α we take the operator

$$\frac{\partial}{\partial t} - \Delta - \alpha \Delta^2$$

and solve the direct problem

$$\frac{\partial u_\alpha}{\partial t} - \Delta u_\alpha - \alpha \Delta^2 u_\alpha = 0, \ x \in D, \ t < T, \ \alpha > 0;$$
$$u_\alpha(x, T) = \psi(x);$$
$$u_\alpha(x, t) = 0 \text{ for } x \in S, \ 0 < t \leqslant T,$$
$$\Delta u_\alpha = 0 \text{ for } x \in S, \ 0 < t \leqslant T.$$

We then set

$$\varphi(x) = u_\alpha(x, 0).$$

It should be pointed out that u_α does not converge in the usual sense as $\alpha \to 0$.

This quasiinversion method can be applied to a broader class of problems dealing with evolution equations (see [104]).

CHAPTER II

THE REGULARIZATION METHOD

In Chapter I, we examined the situation in which the class of possible solutions of equation (1.0.1) is a compact set. However, for a number of applied problems, this class F is not compact and the changes in the right-hand member of the equation

$$Az = u \tag{2.0.1}$$

that are associated with its approximate nature can take u outside the set AF. We shall call such problems **genuinely ill-posed problems**. A new approach to the solution of ill-posed problems that was developed in [156–161] enables us, in the case of genuinely ill-posed problems, to construct approximate solutions of equation (2.0.1) that are stable under small changes in the initial data. This approach is based on the fundamental concept of a regularizing operator [157].

§1. The concept of a regularizing operator.

1. Suppose that the operator A in equation (2.0.1) is such that its inverse A^{-1} is not continuous on the set AF and the set F of possible solutions is not compact.

≠ exact solu.

If the right-hand member of the equation is an element $u_\delta \in U$ that differs from the exact right-hand member u_T by no more than δ, that is, if $\rho_U(u_\delta, u_T) \leqslant \delta$, it is obvious that the approximate solution z_δ of equation (2.0.1) cannot be defined as the exact solution of this equation with approximate right-hand member $u = u_\delta$, that is, according to the formula

$$z_\delta = A^{-1} u_\delta.$$

The numerical parameter δ characterizes the error in the right-hand member of equation (2.0.1). Therefore, it is natural to define z_δ with the aid of an operator depending on a parameter having a value chosen in accordance with the error δ in the initial data u_δ. Specifically, as $\delta \to 0$, that is, as the right-hand member u_δ of equation (2.0.1) approaches (in the metric of the space U) the exact value u_T, the approximate solution z_δ must approach (in the metric of the space F) the exact solution z_T that we are seeking for the equation

$$Az = u_T.$$

Suppose that the elements $z_T \in F$ and $u_T \in U$ are connected by $Az_T = u_T$.

Definition 1. An operator $R(u, \delta)$ is said to be a **regularizing operator** for the equation $Az = u$ in a neighborhood of $u = u_T$ if

1) there exists a positive number δ_1 such that the operator $R(u, \delta)$ is defined for every δ in $0 \leqslant \delta \leqslant \delta_1$ and every $u_\delta \in U$ such that

$$\rho_U(u_T, u_\delta) \leqslant \delta;$$

and

2) for every $\epsilon > 0$, there exists a $\delta_0 = \delta_0(\epsilon, u_T) \leqslant \delta_1$ such that the inequality

$$\rho_U(u_\delta, u_T) \leqslant \delta \leqslant \delta_0$$

implies the inequality

$$\rho_F(z_\delta, z_T) \leqslant \epsilon,$$

where $z_\delta = R(u_\delta, \delta)$.

Remark 1. This definition does not assume uniqueness of the operator R, and z_δ denotes any element in the set $\{R(u_\delta, \delta)\}$.

In many cases, it is more convenient to use another definition of a regularizing operator, one that subsumes the definition just given.

Definition 2. An operator $R(u, \alpha)$ depending on a parameter α is called a regularizing operator for the equation $Az = u$ in a neighborhood of $u = u_T$ if

1) there exists a positive number δ_1 such that the operator $R(u, \alpha)$ is defined for every $\alpha > 0$ and every u in U for which

$$\rho_U(u, u_T) \leqslant \delta \leqslant \delta_1$$

and

2) there exists a function $\alpha = \alpha(\delta)$ of δ such that, for every $\epsilon > 0$, there exists a number $\delta(\epsilon) \leqslant \delta_1$ such that the inclusion $u_\delta \in U$ and the inequality

$$\rho_U(u_T, u_\delta) \leqslant \delta(\epsilon)$$

imply

$$\rho_F(z_T, z_\alpha) \leqslant \epsilon,$$

where $z_\alpha = R(u_\delta, \alpha(\delta))$.

Again, there is no assumption of uniqueness of the operator $R(u_\delta, \alpha(\delta))$. We should point out that here the function $\alpha = \alpha(\delta)$ also depends on u_δ.*

---

*Dependence of the parameter α on u_δ implies that it is also dependent on u_T and hence on z_T since $Az_T = u_T$.

2. If $\rho_U(u_T, u_\delta) \leqslant \delta$, we know from [156, 157] that we can take for an approximate solution of equation (2.0.1) with approximately known right-hand member u_δ the element $z_\alpha = R(u_\delta, \alpha)$ obtained with the aid of the regularizing operator $R(u, \alpha)$, where $\alpha = \alpha(\delta, u_\delta)$ in accordance with the error in the initial data u_δ. This solution is called a **regularized solution** of equation (2.0.1). The numerical parameter α is called the **regularization parameter**. Obviously, every regularizing operator defines a stable method of approximate construction of the solution of equations (2.0.1) provided the choice for α is consistent with the accuracy δ of the initial data ($\alpha = \alpha(\delta)$). If we know that $\rho_U(u_T, u_\delta) \leqslant \delta$, we can, by the definition of a regularizing operator,* choose the value of the regularization parameter $\alpha = \alpha(\delta)$ in such a way that, as $\delta \to 0$, the regularized solution $z_\alpha = R(u_\delta, \alpha(\delta))$ approaches (in the metric of F) the sought exact solution z_T, that is, $\rho_F(z_T, z_{\alpha(\delta)}) \to 0$. This justifies taking as approximate solution of equation (2.0.1) the regularized solution.

Thus, the problem of finding an approximate solution of equation (2.0.1) that is stable under small changes in the right-hand member reduces

a) to finding regularizing operators,

b) to determining the regularization parameter α from supplementary information pertaining to the problem, for example, pertaining to the size of the error in the right-hand member u_δ.

This method of constructing approximate solutions is called the **regularization method.**

3. Out of all the operators $R(u, \alpha)$ from U into F that depend on the parameter α and that are defined for every $u \in U$ and every positive α, we need to single out the operators that are continuous

*In this definition, u_T is assumed to be fixed and for this reason the dependence of R on u_T is not explicitly emphasized. The only information assumed regarding u_T appears in the definition of δ. When we have supplementary information regarding u_T, we naturally use other definitions of a regularizing operator that take this information into account.

with respect to u. For these, we can give sufficient conditions for belonging to the set of regularizing operators of equation (2.0.1). This is an immediate consequence of the

Theorem. *Let A denote an operator from F into U and let $\bar{R}(u, \alpha)$ denote an operator from U into F that is defined for every element u of U and every positive α and that is continuous with respect to u. If*

$$\lim_{\alpha \to 0} \bar{R}(Az, \alpha) = z$$

for every element z of F, then the operator $\bar{R}(u, \alpha)$ is a regularizing operator for the equation

$$Az = u.$$

Proof. It will be sufficient to show that the operator $\bar{R}(u, \alpha)$ possesses property 2) of definition 2 (see p. 46).

Let z_T and u_T denote fixed elements of F and U respectively such that $Az_T = u_T$. Let δ denote a fixed positive number. Then, for every element u_δ of U such that

$$\rho_U(u_\delta, u_T) \leqslant \delta,$$

we have

$$\rho_F(\bar{R}(u_\delta, \alpha), z_T) \leqslant \rho_F(\bar{R}(u_\delta, \alpha), \bar{R}(u_T, \alpha)) + \rho_F(\bar{R}(u_T, \alpha), z_T). \tag{2.1.1}$$

Since the operator $\bar{R}(u, \alpha)$ is continuous with respect to u at the "point" u_T, it follows that, for sufficiently small positive $\delta \leqslant \delta_1$, the inequality

$$\rho_U(u_\delta, u_T) \leqslant \delta \tag{2.1.2}$$

implies the inequality

$$\rho_F(\bar{R}(u_\delta, \alpha), \bar{R}(u_T, \alpha)) \leqslant \omega(\delta), \tag{2.1.3}$$

where $\omega(\delta) \to 0$ as $\delta \to 0$.

Since

$$\lim_{\alpha \to 0} \overline{R}(Az_T, \alpha) = \lim_{\alpha \to 0} \overline{R}(u_T, \alpha) = z_T,$$

there exists for every positive δ an $\alpha_1 = \alpha_1(\delta, z_T)$ such that, for $\alpha \leqslant \alpha_1$,

$$\rho_F(\overline{R}(u_T, \alpha), z_T) \leqslant \omega(\delta). \tag{2.1.4}$$

It follows from inequalities (2.1.1), (2.1.3), and (2.1.4) that, for every $\delta \leqslant \delta_1$ and $\alpha \leqslant \alpha_1$,

$$\rho_F(\overline{R}(u_\delta, \alpha), z_T) \leqslant 2\omega(\delta). \tag{2.1.5}$$

Since $\omega(\delta) \to 0$ as $\delta \to 0$, there exists for every $\epsilon > 0$ a $\delta(\epsilon)$ such that, for $\delta \leqslant \delta(\epsilon) \leqslant \delta_1$ and $\alpha = \alpha_1(\delta, z_T)$, the inequalities (2.1.2) and (2.1.5) imply

$$\rho_F(\overline{R}(u_\delta, \alpha), z_T) \leqslant \epsilon.$$

This completes the proof of the theorem.

Remark 2. Operators $R(u, \alpha)$ which depend on the numerical parameter α have been examined in mathematical literature in connection with existence proofs for solutions of various problems (including ill-posed ones) and also in generalizations of convergence of series. If these operators are continuous with respect to u they define, with a suitable correspondence between α and δ, methods for finding approximate solutions.

§2. Methods of constructing regularizing operators.

A method is presented in [156–161] for constructing regularizing operators for equations (2.0.1), which is based on a variational principle. We shall describe it in the present section (see also [211]). We shall assume that the equation $Az = u_T$ has a unique solution z_T.

1. Let $\Omega[z]$ denote a continuous nonnegative functional defined on a subset F_1 of F that is everywhere dense in F. Suppose that

a) z_T belongs to the domain of definition of $\Omega[z]$,

b) for every positive number d, the set of elements z of F_1 for which $\Omega[z] \leqslant d$ is a compact subset of F_1.

We shall refer to functionals $\Omega[z]$ possessing these properties as **stabilizing functionals.**

Suppose that we know that the deviation of the right-hand member u_δ from the exact value u_T does not exceed δ, that is, $\rho_U(u_\delta, u_T) \leqslant \delta$. It then is natural to seek an approximate solution in the class Q_δ of elements z such that $\rho_U(Az, u_\delta) \leqslant \delta$. This Q_δ is the set of possible solutions. However, we cannot take an arbitrary element z_δ of Q_δ as the approximate solution of equation (2.0.1) with approximate right-hand member $u = u_\delta$ because such a "solution" will not in general be continuous with respect to δ. Thus, the set Q_δ is too broad. We need a principle for selecting the possible solutions that ensures that we obtain as approximate solution an element (or possibly more than one element) of Q_δ that depends continuously on δ. For such a principle we may take the above-mentioned variational principle, which is also applicable to the construction of approximations to a quasisolution if one exists.

Suppose that $\Omega[z]$ is a stabilizing functional defined on a subset (proper or otherwise) F_1 of the set F. We shall consider only those elements of Q_δ on which the functional $\Omega[z]$ is defined; that is, we shall consider only elements of the set $F_{1,\delta} = Q_\delta \cap F_1$. Among the elements of this set, let us find the one (or ones) that will minimize the functional $\Omega[z]$ on $F_{1,\delta}$. Let z_δ denote such an element.* It can be regarded as the result of applying to the right-hand member $u = u_\delta$ of equation (2.0.1) an operator $\widetilde{R}$ depending on the parameter δ:

$$z_\delta = \widetilde{R}(u_\delta, \delta).$$

*The existence of such an element will be proven below.

2. We shall show that the operator $\widetilde{R}(u, \delta)$ is a regularizing operator for equation (2.0.1) and hence that the element

$$z_\delta = \widetilde{R}(u_\delta, \delta)$$

can be taken as approximate solution of equation (2.0.1).

Let us show first of all that the operator $\widetilde{R}(u, \delta)$ is defined for every $\delta > 0$ and every $u_\delta \in U$ such that

$$\rho_U(u_\delta, u_T) \leqslant \delta.$$

Since $\Omega[z]$ is a nonnegative functional, there exists

$$\inf_{z \in F_{1,\delta}} \Omega[z] = \Omega_0.$$

Let $\{z_n\}$ denote a minimizing sequence for $\Omega[z]$, that is, one such that

$$\lim_{n \to \infty} \Omega[z_n] = \Omega_0.$$

We may assume without loss of generality that, for every $n > 1$,

$$\Omega[z_n] \leqslant \Omega[z_{n-1}] \leqslant \ldots \leqslant \Omega[z_1].$$

Thus, the sequence $\{z_n\}$ belongs to the compact (in itself) set of elements z of F_1 such that

$$\Omega[z] \leqslant \Omega[z_1].$$

Consequently, this sequence has a convergent subsequence $\{z_{n_k}\}$. Let us define

$$\lim_{n_k \to \infty} z_{n_k} = z_\delta.$$

The compactness of the sequence $\{z_{n_k}\}$ on F_1 implies that z_δ

belongs to the set F_1, so that the functional $\Omega[z]$ is defined on it.*

The continuity of the functional $\Omega[z]$ at the element z_δ implies that

$$\rho_U(Az_{\delta_{n_k}}, u_{\delta_{n_k}}) \leqslant \delta_{n_k}.$$

But since

$$\lim_{n_k \to \infty} \Omega[z_{n_k}] = \Omega_0,$$

we have

$$\Omega[z_\delta] = \Omega_0 = \inf_{z \in F_{1,\delta}} \Omega[z].$$

Thus, property 1) of the definition of a regularizing operator is proven for $\widetilde{R}(u, \delta)$. Let us now show that it possesses property 2) of the definition.

Since the element z_δ minimizes the functional $\Omega[z]$ on the set $F_{1,\delta}$ and $z_T \in F_{1,\delta}$, it is obvious that

$$\Omega[z_\delta] \leqslant \Omega[z_T].$$

Thus, the element z_δ belongs to the set

$$F_T \equiv \{z;\ \Omega[z] \leqslant \Omega[z_T]\},$$

which is compact on F_1.

Suppose that we are given a sequence $\{u_n\}$ such that $\rho_U(u_T, u_n) \leqslant \delta_n$, where $\{\delta_n\}$ is a sequence of positive numbers that converges to 0; that is $\delta_n \to 0$ as $n \to \infty$. For every δ_n, the set Q_{δ_n} is defined. Let us now define

$$F_{1,\delta_n} = Q_{\delta_n} \cap F_1.$$

---

*One can easily see that $z_\delta \in F_{1,\delta}$.

Each of the sets F_{1,δ_n} has, by virtue of what we have just proven, an element z_{δ_n} minimizing the functional $\Omega[z]$ on that set. Thus, corresponding to the sequence of numbers $\{\delta_n\}$ is a sequence of elements $\{z_{\delta_n}\}$ belonging to the set F_T, which is compact on F_1. Consequently, the sequence $\{z_{\delta_n}\}$ has a convergent (in the metric of F) subsequence

$$\{z_{\delta_{n_k}}\}.$$

Let us define

$$\tilde{z} = \lim_{n_k \to \infty} z_{\delta_{n_k}}.$$

Since $z_{\delta_n} \in F_{1,\delta_n} \subset Q_{\delta_n}$, it follows that, for every element $z_{\delta_{n_k}}$ of this subsequence,

$$\rho_U(Az_{\delta_{n_k}}, u_{\delta_{n_k}}) \leqslant \delta_{n_k}.$$

Taking the limit as $n_k \to \infty$ and using the continuity of the operator A, we obtain

$$\rho_U(A\tilde{z}, u_T) = 0.$$

Consequently, $A\tilde{z} = u_T$. The uniqueness of the solution of equation (2.0.1) with right-hand member $u = u_T$ implies that $\tilde{z} = z_T$. Thus,

$$\lim_{n_k \to \infty} z_{\delta_{n_k}} = z_T.$$

This will be the case for every convergent subsequence of the sequence $\{z_{\delta_n}\}$. It follows that, for every sequence $\{\delta_n\}$ of positive numbers δ_n that converges to zero, the corresponding sequence $\{z_{\delta_n}\}$ converges (in the metric of the space F) to the element z_T.

This proves that $\tilde{R}(u, \delta)$ possesses property 2) of the definition of a regularizing operator, so that it is a regularizing operator for equation (2.0.1).

Remark. If the equation $Az = u_T$ has more than one solution, this method can still be used to construct a regularizing operator. In this case, every convergent subsequence $\{z_{\delta_n}\}$ converges to some solution of equation (2.0.1) with right-hand member $u = u_T$ although different subsequences may converge to different solutions.

3. Thus, with this approach, the problem of finding an approximate solution of equation (2.0.1) with approximate right-hand member consists in minimizing the functional $\Omega[z]$ on the set

$$F_{1,\delta} \equiv Q_\delta \cap F_1,$$

where

$$Q_\delta \equiv \{z;\ \rho_U(Az, u_\delta) \leqslant \delta\}.$$

It should be pointed out that numerical solution of this last problem on a computer is often difficult.

Define

$$\Omega_0 = \inf_{z \in F_1} \Omega[z]$$

and let M_0 denote the set of all elements z of F_1 such that $\Omega[z] = \Omega_0$.

Let us assume for simplicity that M_0 consists of a single element z_0.* There are two possibilities:

1) The sets M_0 and $F_{1,\delta}$ have one or more members in common;

2) the sets M_0 and $F_{1,\delta}$ are disjoint.

In the first case, we take the element z_0 as the solution of the variational problem of minimizing the functional $\Omega[z]$ on the set

*If the set M has more than one element, stability of the solution should be understood in the sense of continuity of multiple-valued mappings.

$F_{1,\delta}$. This solution is stable under small changes in u_δ. To see this, note that, for every $\epsilon > 0$, the solution belongs to the set D_ϵ (compact on F_1) of elements z such that $\Omega[z] \leqslant \Omega_0 + \epsilon$. The operator A is continuous and one-to-one. The lemma of §1 of chapter I then tells us that the inverse mapping is also continuous.

In the second case, $\rho_U(u_\delta, AM_0) > \delta$ and, for a broad class of cases, we can reduce the problem of minimizing the functional $\Omega[z]$ on the set $F_{1,\delta}$ to the classical problem of finding a conditional extremum of the functional $\Omega[z]$. This problem is much more amenable to numerical solution on a computer. The reduction can be made with the aid of the lemma to be given below.

We shall say that the functional $\Omega[z]$ is **quasimonotonic*** if, for every element z_0 of F_1 that does not belong to the set M_0, every neighborhood of it includes an element z_1 of F_1 such that $\Omega[z_1] < \Omega[z_0]$.

Lemma. The greatest lower bound of a quasimonotonic functional $\Omega[z]$ on a set $F_{1,\delta}$ for which $M_0 \cap F_{1,\delta}$ is empty is attained for an element z_δ for which $\rho_U(Az_\delta, u_\delta) = \delta$.

Proof. Let us suppose that

$$\inf_{z \in F_{1,\delta}} \Omega[z]$$

is attained at an element z_δ of $F_{1,\delta}$ for which

$$\rho_U(Az_\delta, u_\delta) = \beta < \delta.$$

By virtue of the restriction imposed on the size of the error δ, we have $z_\delta \notin M_0$. The continuity of the operator A on F implies the existence of a neighborhood $O(z, z_\delta)$ of the element z_δ all elements of which satisfy the inequality

$$\rho_U(Az, Az_\delta) < \frac{\delta - \beta}{2}.$$

*M. M. Lavrent'yev and other authors have used this term but with a different meaning.

For every z in that neighborhood, we have

$$\rho_U(Az, u_\delta) < \delta,$$

since

$$\rho_U(Az, u_\delta) \leqslant \rho_U(Az, Az_\delta) + \rho_U(Az_\delta, u_\delta) <$$
$$< \frac{\delta - \beta}{2} + \beta = \frac{\delta + \beta}{2} < \delta.$$

Consequently,

$$O(z, z_\delta) \subset Q_\delta.$$

The quasimonotonicity of the stabilizing functional $\Omega[z]$ implies that the neighborhood $O(z, z_\delta)$ includes an element $z_{\delta,1}$ of F_1 such that

$$\Omega[z_{\delta,1}] < \Omega[z_\delta]. \tag{2.2.1}$$

Since the element $z_{\delta,1}$ belongs to the sets Q_δ and F_1 and hence to the set $F_{1,\delta}$, inequality (2.2.1) is in contradiction with the assumption that the functional $\Omega[z]$ attains its greatest lower bound on the set $F_{1,\delta}$ at z_δ. This completes the proof of the lemma.

We can use this lemma to solve the problem of minimizing the functional $\Omega[z]$ not on the set $F_{1,\delta}$ but on the set F_1 under the condition that the element z that we are seeking satisfies the equation

$$\rho_U(Az, u_\delta) = \delta.$$

This is a conditional extremum problem. Let us solve it by the method of undetermined Lagrange multipliers; that is, let us minimize the functional

$$M^\alpha[z, u_\delta] = \rho_U^2(Az, u_\delta) + \alpha\Omega[z], \tag{2.2.2}$$

where the numerical parameter α is determined from the condition $\rho_U(Az_\alpha, u_\delta) = \delta$, where z_α is an element at which the functional $M^\alpha[z, u_\delta]$ attains its greatest lower bound.

If Lagrange's method is realizable, that is, if there exists an α such that $\rho_U(Az_\alpha, u_a) = \delta$, then the original variational problem is equivalent to the problem of minimizing the functional $M^\alpha[z, u_\delta]$. Specifically, if α is chosen in such a way that $\rho_U(Az_\alpha, u_\delta) = \delta$, then the solution z_δ of the original variational problem also minimizes the functional $M^\alpha[z, u_\delta]$. Conversely, if z_α minimizes the functional $M^\alpha[z, u_\delta]$ under the condition $\rho_U(Az, u_\delta) = \delta$, then the minimum of the functional $\Omega[z]$ is attained at the same element z_α. The question of sufficient conditions for realizability of Lagrange's method is examined in greater detail in section 6 of the present chapter.*

The element z_α can be regarded as the result of applying to the right-hand member $u = u_\delta$ of equation (2.0.1) some operator R_1 depending on the parameter α:

$$z_\alpha = R_1(u_\delta, \alpha),$$

where $\alpha = \alpha(\delta)$ depends on the discrepancy (see section 6).

Thus, as approximate solution of equation (2.0.1), we take the solution of a *different* problem (the problem of minimizing the functional $M^\alpha[z, u]$) that is close to the original one for small values of the error in the right-hand member u_δ.**

4. It should be mentioned that it is possible to consider the functional $M^\alpha[z, u]$ without connecting it with the question of a conditional extremum of the functional $\Omega[z]$ and to seek an element z_α minimizing it on the set F_1. The problem then arises of finding the regularization parameter α in the form of a function depending on δ and on the other parameters of the problem ($\alpha = \alpha(\delta)$) such that the operation $R_1(u, \alpha(\delta))$ defining the element

*It is possible to construct an example in which Lagrange's method for solving conditional extremum problems is not realizable; that is, α may not be defined by the condition $\rho_U(Az_\alpha, u_\delta) = \delta$.

**See Theorem 2 of § 3 of the present chapter.

$$z_\alpha = R_1(u, \alpha(\delta)),$$

will be a regularizing operator for equation (2.0.1). Under certain conditions, such a function exists and can be found, for example, from the relationship $\rho_U(Az_\alpha, u_\delta) = \delta$. We shall see in §3 that a set of such functions $\alpha(\delta)$ exists. The possibility of using different functions $\alpha(\delta)$ in the regularizing operator $R_1(u, \alpha(\delta))$ is connected with the nonuniqueness of regularizing operators for equation (2.0.1). We shall look in greater detail in §§ 3 and 6 at ways of determining the regularization parameter.

In connection with what has been said, the problem of finding the optimal value (in a sense determined in advance) of the regularization parameter $\alpha = \alpha(\delta)$ is a natural one. For certain operators A of the convolution type, this problem will be examined in Chapter V.

5. It should be pointed out that, whereas the original problem (2.0.1) does not have the property of stability, we shall show in §3 that the problem of minimizing the functional $M^\alpha[z, u]$ is stable under small changes in the right-hand member u. This stability has been attained by narrowing the class of possible solutions through the introduction of the functional $\Omega[z]$ with the properties described above. Thus, Ω plays a stabilizing role. For this reason, it is called the **stabilizing functional** for the problem (2.0.1) or simply the **stabilizer.**

The choice of stabilizing functional $\Omega[z]$ is often prompted by the nature of the problem. However, in a number of cases, more than one choice is possible. We shall call the functional $M^\alpha[z, u]$ a **smoothing functional.**

6. We have described a method for constructing regularizing operators that is based on a variational principle. In what follows, we shall call it the variational method of constructing regularizing operators. There are other ways of doing this. A method based on the use of the spectrum of the operator A is described, for example, in [16, 18, 19] (see also [104, 185–187]). For operators A of convolution type, a method is given in Chapter IV for constructing a regularizing operator by using Fourier, Laplace, Mellin, and other integral transformations. Regularizing operators

are also examined in [5–7] and [149] for equations of convolution type.

7. Approximate solutions of the equation $Az = u$ that are stable under small changes in the initial data can also be constructed by the method of iterations (see [17, 91, 92, 99, 112, 117, 151, 153]), taking $z_n = R(u, z_{n-1}, \ldots, z_{n-k})$, where $k < n$. For these solutions to be stable under small changes in the initial data, the iteration number $n = n(\delta)$ yielding z_n (taken as the approximate solution) must be compatible with the size of the error in the initial data.

Sometimes, we can obtain an *a priori* estimate of the deviation of z_n from z_T (see [99, 117])

$$\rho_F(z_n, z_T) \leqslant B(\delta, n)$$

and then, by minimizing $B(\delta, n)$, find $n(\delta)$. In some cases, $n(\delta)$ can be found from the discrepancy.

8. We have examined in detail methods for constructing approximate solutions of ill-posed problems of the form (2.0.1) that are stable under small changes in the initial information when the right-hand member of (2.0.1) is inexactly given but the operator A is assumed to be known exactly. Let us look at the problem of constructing a solution of equation (2.0.1) in cases in which both the right-hand member of the equation and the operator A are given approximately [46, 47].

Let z_T denote the exact solution of the ill-posed problem (2.0.1) with right-hand member $u = u_T$. Suppose that $z_T \in F$ and $u_T \in U$. Let A denote a continuous one-to-one operator from F into U. Suppose that we are given not the exact initial data $\{A, u_T\}$ of the problem (2.0.1) but a two-parameter family of approximate initial data $\{A_h, u_\delta\}$, the closeness of which to the exact data is characterized by a pair $\gamma = (h, \delta)$ of nonnegative numbers h and δ. Then, it is only a question of finding an approximate solution (i.e., one close to z_T) of the equation

$$A_h z = u_\delta. \tag{2.2.3}$$

Let $\Omega[z]$ denote a quasimonotonic stabilizing functional defined on a subset F_1 of F. Let us suppose that the numbers δ and h characterize the approximate nature of the initial data $\{A_h, u_\delta\}$ in the following sense:

$$\rho_U(u_T, u_\delta) \leqslant \delta, \; h = \sup_{\substack{z \in F_1 \\ \Omega[z] \neq 0}} \frac{\rho_U(Az, A_h z)}{\{\Omega[z]\}^{1/2}}, \; h < \infty.$$

Let us suppose that, for arbitrary $h \geqslant 0$, the operator A_h maps F continuously and one-to-one onto U and that $A_0 = A$. Following the idea of the method of regularization expounded in §2, we can formulate the above-mentioned problem of finding an approximate solution of equation (2.2.3) that is stable under small changes in the initial information as follows: Out of all elements z belonging to F_1 such that

$$\rho_U^2(A_h z, u_\delta) \leqslant \delta^2 + h^2\Omega[z],$$

find an element z_γ giving the functional $\Omega[z]$ its greatest lower bound on the set F_1.

Lemma. *The greatest lower bound of the functional* $\Omega[z]$ *is attained with an element* z_γ *of* F_1 *for which*

$$\rho_U^2(A_h z_\gamma, u_\delta) = \delta^2 + h^2\Omega[z_\gamma].$$

Proof. We proceed in the same manner as in §2. Let us suppose that $\inf_{z \in F_1} \Omega[z]$ is attained with an element z_0 of F_1 for which

$$\rho_U^2(A_h z_0, u_\delta) = \delta_0^2 < \delta^2 + h^2\Omega[z_0] = \Delta_0^2.$$

The continuity of the operator A_h (from F into U) implies that, for every number δ_1 such that $\delta_0 < \delta_1 < \Delta_0$, there exists a neighborhood $O_1(z, z_0)$ of the element z_0 any element z in which satisfies the inequality $\rho_U(A_h z, u_\delta) < \delta_1$. Since the functional $\Omega[z]$ is continuous, there exists a neighborhood $O_2(z, z_0)$ of the element z_0 in which every element z satisfies the inequality

$\delta_1^2 < \delta^2 + h^2\Omega[z]$. Consequently, for all elements z belonging to the intersection of the neighborhoods $O_1(z, z_0)$ and $O_2(z, z_0)$, we have

$$\rho_U^2(A_h z, u_\delta) < \delta^2 + h^2\Omega[z]. \qquad (2.2.4)$$

Since the functional $\Omega[z]$ is quasimonotonic, every neighborhood $O_3(z, z_0)$ of the element z_0 that is contained in the intersection $O_1(z, z_0) \cap O_2(z, z_0)$ includes an element z_1 for which $\Omega[z_1] < \Omega[z_0]$. Since inequality (2.2.4) is satisfied for the element z_1, this means that $\inf_{z \in F_1} \Omega[z]$ is not attained with the element z_0. This contradiction with the assumption proves the lemma.

According to this lemma, the problem of minimizing the functional $\Omega[z]$ on the set F_1 under the condition $\rho_U^2(A_h z, u_\delta) \leqslant \delta^2 + h^2\Omega[z]$ reduces to the problem of minimizing $\Omega[z]$ under the condition $\rho_U^2(A_h z, u_\delta) = \delta^2 + h^2\Omega[z]$ and is solved in accordance with the method described in §2 by minimizing the corresponding smoothing functional

$$M^\alpha[z, u_\delta, A_h] = \rho_U^2(A_h z, u_\delta) + (\alpha - h^2)\Omega[z].$$

Here, the parameter α can be determined from the discrepancy by the condition

$$\rho_U^2(A_h z_\gamma^\alpha, u_\delta) - h^2\Omega[z_\gamma^\alpha] = \delta^2.$$

The operator $R_2(u_\delta, A_h, h, \delta)$ which assigns to the initial data of the problem $\{A_h, u_\delta\}$ an element z_γ, where $\gamma = (h, \delta)$, minimizing the functional $\Omega[z]$ on the set of elements of F_1 for which $\rho_U^2(A_h z, u_\delta) = \delta^2 + h^2\Omega[z]$ is a regularizing operator for the problem (2.2.3).

It should be pointed out that the element z_γ is not necessarily unique.

§3. The construction of regularizing operators by minimization of a smoothing functional.

We can construct the regularizing operator for equation (2.0.1) by minimizing the smoothing functional $M^{\alpha}[z, u]$. The parameter α is then determined from the discrepancy by the condition

$$\rho_U(Az_{\alpha}, u_{\delta}) = \delta.$$

This method of constructing the regularizing operator is, as we saw in §2, equivalent to the method of constructing it by minimizing the functional $\Omega[z]$ on the set of elements z for which

$$\rho_U(Az, u_{\delta}) \leqslant \delta.$$

As we noted in §2, the smoothing functional $M^{\alpha}[z, u]$ can be defined formally without connecting it with the variational problem of finding a conditional extremum of the functional $\Omega[z]$ and the regularizing operator can be constructed by solving the problem of minimizing the functional $M^{\alpha}[z, u]$. Here, we need to take for α the corresponding function of δ in accordance with definition 2 of a regularizing operator. In the present section, we shall show that it is possible to obtain in this way a broad class of regularizing operators.

We point out here that the regularization method described below for finding an approximate solution of equation (2.0.1) and its justification can be applied without any change to the approximate determination of a quasisolution of the same equation.

1. Suppose that the set F of possible solutions of equation (2.0.1) is a metric space and that $\Omega[z]$ is a stabilizing functional defined on a set $F_1 \subset F$. Then, we have

Theorem 1. *Let A denote a continuous operator from F into U. For every element u of U and every positive parameter α, there exists an element $z_{\alpha} \in F_1$ for which the functional*

$$M^{\alpha}[z, u] = \rho_U^2(Az, u) + \alpha\Omega[z]$$

attains its greatest lower bound:

$$\inf_{z \in F_1} M^\alpha [z, u] = M^\alpha [z_\alpha, u].$$

Proof. Since $M^\alpha \geqslant 0$ for every $z \in F_1$, the quantity $\inf M^\alpha = M_0{}^\alpha$ exists, the infimum being over all admissible elements of F_1. There exists a minimizing sequence $\{z_n{}^\alpha\}$ of elements of F_1 such that $\lim_{n \to \infty} M_n^\alpha = M_0^\alpha$, where $M_n^\alpha = M^\alpha [z_n^\alpha, u]$. Obviously, we may assume that, for every n,

$$M_{n+1}^\alpha \leqslant M_n^\alpha \leqslant M_1^\alpha.$$

Then, for every n and every fixed $\alpha > 0$,

$$\Omega [z_n^\alpha] \leqslant \frac{1}{\alpha} M_1^\alpha = Q.$$

Thus, the sequence $\{z_n^\alpha\}$ belongs to the set of elements z of F_1 for which $\Omega[z] \leqslant Q$. Since this set is a compact subset of F_1, the sequence $\{z_n^\alpha\}$ has a subsequence $\{z_{n_k}^\alpha\}$ that converges (with respect to the metric of F) to some element z_α of F_1. The continuity of the operator A implies

$$\inf_{z \in F_1} M^\alpha [z, u] = \lim_{n \to \infty} M^\alpha [z_n^\alpha, u] = \lim_{n_k \to \infty} M^\alpha [z_{n_k}^\alpha, u] =$$

$$= \lim_{n_k \to \infty} \{\rho_U^2 (A z_{n_k}^\alpha, u) + \alpha \Omega [z_{n_k}^\alpha]\} = \rho_U^2 (A z_\alpha, u) + \alpha \Omega [z_\alpha].$$

This completes the proof of the theorem.

Thus, an operator $R_1(u, \alpha)$ into F_1 is defined on the set of pairs (u, α), where $u \in U$ and $\alpha > 0$, so that the element

$$z_\alpha = R_1 (u, \alpha)$$

minimizes the functional $M^\alpha [z, u]$.

Sufficient conditions for uniqueness of the element z_α can be exhibited. This will be the case, for example, if the operator A is linear, the set F is a Hilbert space, and $\Omega[z]$ is a quadratic stabilizing functional.

To see this, let us suppose that there exist two elements $z_\alpha^{(1)}$ and $z_\alpha^{(2)}$ for which the functional $M^\alpha[z, u]$ attains its greatest lower bound. Consider elements of the space F_1 located on the line segment (in the space F) connecting $z_\alpha^{(1)}$ and $z_\alpha^{(2)}$:

$$z = z_\alpha^{(1)} + \beta\,(z_\alpha^{(2)} - z_\alpha^{(1)}).$$

On the elements of this line, the functional $M^\alpha[z, u]$ is a nonnegative quadratic function of β. Consequently, it cannot attain its least value at two distinct values of β. For a nonlinear operator A, the element z_α may not be unique.

2. We need to show that the operator $R_1(u, \alpha)$ is a regularizing operator for equation (2.0.1).

Let us denote by T_{δ_1} the class of functions that are nonnegative, nondecreasing, and continuous on an interval $[0, \delta_1]$.

Theorem 2. *Let z_T denote a solution of equation* (2.0.1) *with right-hand member $u = u_T$; that is, $Az_T = u_T$. Then, for any positive number ϵ and any functions $\beta_1(\delta)$ and $\beta_2(\delta)$ in the class T_{δ_1} such that $\beta_2(0) = 0$ and $\delta^2/\beta_1(\delta) \leqslant \beta_2(\delta)$, there exists a number $\delta_0 = \delta_0(\epsilon, \beta_1, \beta_2) \leqslant \delta_1$ such that for $\tilde{u} \in U$ and $\delta \leqslant \delta_0$ the inequality $\rho_U(\tilde{u}, u_T) \leqslant \delta$ implies the inequality $\rho_F(z_T, \tilde{z}_\alpha) \leqslant \epsilon$, where $\tilde{z}_\alpha = R_1(\tilde{u}, \alpha)$ for all α satisfying the inequalities*

$$\frac{\delta^2}{\beta_1(\delta)} \leqslant \alpha \leqslant \beta_2(\delta).$$

Proof. Since the functional $M^\alpha[z, \tilde{u}]$ attains its minimum when $z = \tilde{z}_\alpha$, we have

$$M^\alpha[\tilde{z}_\alpha, \tilde{u}] \leqslant M^\alpha[z_T, \tilde{u}].$$

Therefore,

$$\begin{aligned}\alpha\Omega[\tilde{z}_\alpha] \leqslant M^\alpha[\tilde{z}_\alpha, \tilde{u}] \leqslant M^\alpha[z_T, \tilde{u}] &= \\ = \rho_U^2(Az_T, \tilde{u}) + \alpha\Omega[z_T] = \rho_U^2(u_T, \tilde{u}) + \alpha\Omega[z_T] &\leqslant \\ \leqslant \delta^2 + \alpha\Omega[z_T] = \alpha\left\{\frac{\delta^2}{\alpha} + \Omega[z_T]\right\}&.\end{aligned}$$

The inequality $\delta^2/\beta_1(\delta) \leqslant \alpha$ implies that $\frac{\delta^2}{\alpha} \leqslant \beta_1(\delta) \leqslant \beta_1(\delta_1)$ and $\frac{\delta^2}{\alpha} + \Omega[z_T] \leqslant \beta_1(\delta_1) + \Omega[z_T] = H_0$. Thus,

$$\Omega[\tilde{z}_\alpha] \leqslant H_0 \text{ and } \Omega[z_T] \leqslant H_0.$$

Consequently, the elements z_T and $\tilde{z}_\alpha$ belong to the set F_{H_0} of elements z of F_1 such that

$$\Omega[z] \leqslant H_0.$$

This set is compact in F_1. Let U_{H_0} denote the image of the set F_{H_0} under the mapping $u = Az$. Since the operator A is continuous, the mapping $F_{H_0} \rightarrow U_{H_0}$ is continuous and hence the solution of the equation $Az = u$ is unique for every $u \in U_{H_0}$ and the set F_{H_0} is compact. Therefore, by the lemma of Chapter I, the inverse mapping $U_{H_0} \rightarrow F_{H_0}$ is also continuous (in the metric of F). This means that for arbitrary $\epsilon > 0$ there exists a number $\gamma(\epsilon) > 0$ such that the inequality

$$\rho_U(u_1, u_2) \leqslant \gamma(\epsilon), \quad u_1, u_2 \in U_{H_0}$$

implies the inequality

$$\rho_F(z_1, z_2) \leqslant \epsilon$$

if $u_1 = Az_1$ and $u_2 = Az_2$.

Furthermore, for $\tilde{u}$ and $\tilde{u}_\alpha = A\tilde{z}_\alpha$, we have

$$\begin{aligned}\rho_U^2(\tilde{u}_\alpha, \tilde{u}) = \rho_U^2(A\tilde{z}_\alpha, \tilde{u}) \leqslant M^2[\tilde{z}_\alpha, \tilde{u}] \leqslant M^\alpha[z_T, \tilde{u}] = \\ = \rho_U^2(Az_T, \tilde{u}) + \alpha\Omega[z_T] = \\ = \rho_U^2(u_T, \tilde{u}) + \alpha\Omega[z_T] \leqslant \delta^2 + \alpha\Omega[z_T].\end{aligned}$$

Using the inequality $\alpha \leqslant \beta_2(\delta)$, we obtain

$$\rho_U(\tilde{u}_\alpha, \tilde{u}) \leqslant \{\delta^2 + \beta_2(\delta)\,\Omega[z_T]\}^{1/2} = \varphi(\delta), \qquad (2.3.1)$$

where $\varphi(\delta)$ is a continuous monotonically increasing function on $[0, \delta_1]$ and $\varphi(0) = 0$. Obviously,

$$\rho_U(\tilde{u}_\alpha, u_T) \leqslant \rho_U(\tilde{u}_\alpha, \tilde{u}) + \rho_U(\tilde{u}, u_T). \qquad (2.3.2)$$

Using inequality (2.3.1) and the inequality

$$\rho_U(\tilde{u}, u_T) \leqslant \delta,$$

we obtain

$$\rho_U(\tilde{u}_\alpha, u_T) \leqslant \delta + \varphi(\delta) = \psi(\delta), \qquad (2.3.3)$$

where $\psi(\delta)$ is a continuous monotonically increasing function on $[0, \delta_1]$ for which $\psi(0) = 0$. Setting $\delta_0 = \psi^{-1}(\gamma(\epsilon))$, where $\psi^{-1}(y)$ is the inverse of the function $y = \psi(\delta)$, and using the continuity of the mapping $U_{H_0} \to F_{H_0}$, we see that, for all α in the closed interval

$$\frac{\delta^2}{\beta_1(\delta)} \leqslant \alpha \leqslant \beta_2(\delta),$$

the inequality

$$\rho_U(\tilde{u}, u_T) \leqslant \delta \leqslant \delta_0$$

implies the inequality

$$\rho_F(z_T, \tilde{z}_\alpha) \leqslant \epsilon.$$

This completes the proof of the theorem. Estimates of the errors in the approximate solutions are examined in [26, 27, 77, 80, 85, 86, 117].

This theorem shows that, when we construct regularizing operators by minimizing the smoothing functional $M^\alpha[z, u]$, the regularization parameter α is a multiple-valued function of the error in the right-hand member δ. This function can be defined not only in terms of the discrepancy, that is, from the condition $\rho_U(Az_\alpha^\delta, u_\delta) = \delta$, as was done in §2, but also in other ways (see §6).

Remark 1. Frequently, restrictions of the form $\varphi_1(x) \leqslant z_T(x) \leqslant \varphi_2(x)$, where $\varphi_1(x)$ and $\varphi_2(x)$ are given functions, are imposed by the nature of the problem on the solution $z_T(x)$ that we are seeking. For example, the solution must be nonnegative ($\varphi_1(x) \equiv 0$). In such cases, we need to take as the space of possible solutions F the space of functions $z(x)$ satisfying those inequalities. Theorems 1 and 2 of the present section remain valid in these cases.

3. Suppose that a subset Φ of a metric space F admits a metrization $\rho_\Phi(z_1, z_2)$ that majorizes the metric $\rho_F(z_1, z_2)$ of the space F; that is, for every pair of elements z_1 and z_2 in Φ,

$$\rho_F(z_1, z_2) \leqslant \rho_\Phi(z_1, z_2).$$

If the sphere

$$\rho_\Phi(z, z_0) \leqslant d$$

(with center at z_0) is compact in F (in the metric of F), then Theorem 1 is valid; that is, there exists an element $z_\alpha \in \Phi$ minimizing the functional

$$M^\alpha[z, \tilde{u}] = \rho_U^2(Az, \tilde{u}) + \alpha\Omega[z], \tag{2.3.4}$$

where

$$\Omega[z] = \rho_\Phi^2(z; 0).$$

If the exact solution z_T of equation (2.0.1) that we are seeking belongs to the set Φ, then the operator $R_2(\tilde{u}, \alpha)$ which provides, for every $\alpha > 0$ and $\tilde{u} \in U$, an element $\tilde{z}_\alpha$ minimizing the functional (2.3.4) is a regularizing element. The proofs of these assertions are analogous to those given in sections 1 and 2.

Thus, if F is the set of continuous functions $z(x)$ on the interval $[a, b]$ with metric

$$\rho_F(z_1, z_2) = \sup_{x \in [a,b]} |z_1(x) - z_2(x)|,$$

then we can take for the set Φ the set C_1 of continuously differentiable functions on that interval with metric

$$\rho_\Phi(z_1, z_2) = \sup_{a \leqslant x \leqslant b} \{|z_1(x) - z_2(x)| + |z_1'(x) - z_2'(x)|\}.$$

By Arzelà's theorem, any sequence $\{z_n(x)\}$ of continuous functions $z_n(x)$ that satisfy the inequality

$$\rho_\Phi(z_n, z_0) \leqslant d$$

contains a subsequence $\{z_{n_k}(x)\}$ that converges uniformly to a continuous function $z_0(x) \in F$, that is, that converges to $z_0(x)$ in the sense of the metric of F. Consequently, the sphere $\{z;\ \rho_\Phi(z, \bar{z}_0) \leqslant d\}$ is compact in C.

As a second example, consider the case in which F is the space of continuous functions $z(x)$ on the interval $[a, b]$ with the C-metric

$$\rho_F(z_1, z_2) = \sup_{a \leqslant x \leqslant b} |z_1(x) - z_2(x)|,$$

and Φ is the space of functions that have square-integrable generalized derivatives up to pth order; that is, Φ is the Sobolev space W_2^p. The metric in W_2^p is defined by

$$\rho_\Phi(z_1, z_2) = \left\{ \int_a^b \sum_{r=0}^{p} q_r(x) \left(\frac{d^r z}{dx^r} \right)^2 dx \right\}^{1/2}, \quad z = z_1 - z_2,$$

where $q_0(x), q_1(x), \ldots, q_{p-1}(x)$ are given nonnegative continuous functions and $q_p(x)$ is a given positive continuous function. It is well known that, for every p, the space W_2^p is a Hilbert space and that a closed ball in it is compact in C. Consequently, if we seek regularized solutions of equation (2.0.1) in the space W_2^p, Theorems 1 and 2 are also valid for them.

Remark 2. Since in this case a regularized solution $z_\alpha(x)$ minimizes the stabilizing functional

$$\Omega[z] = \int_a^b \sum_{r=0}^{p} q_r(x) \left(\frac{d^r z}{dx^r} \right)^2 dx \tag{2.3.5}$$

(under the condition $\rho_U(u, Az) = \delta$), it will obviously be a "smoothest" function (up to order p) for which $\rho_U(Az, u) = \delta$. Thus, in this case, we approximate the sought solution z_T with the aid of the "smoothest" functions (up to order p).

Stabilizers of the form (2.3.5), where $q_r(x) \geqslant 0$ for $r = 0, 1, \ldots, p-1$ and $q_p(x) > 0$ will be called **stabilizers of pth order**. If all the functions $q_r(x)$ are constants, they will be called **stabilizers of pth order with constant coefficients.***

Conditions under which regularizing operators exist are examined in [25].

Remark 3. The results of §§ 1–3 were obtained for equations (2.0.1), in which the operator A is continuous. However, these results can be carried over to equations in which the operator A is closed (see [61, 106–108, 114]).

Remark 4. The regularization method that we have described can also be used to solve well-posed problems of the form $Az = u$, for example, to solve Fredholm integral equations of the second kind.

Various questions associated with ill-posed problems are examined in the following papers [1, 4, 8, 21, 22, 54, 81, 88, 89, 110, 113, 123, 147, 150, 152, 154, 163, 169, 174, 191, 193, 197-200, 204-206, 208, 213, 215, 216, 218, 219].

4. The concepts of quasisolutions and of regularized solutions of the equation $Az = u$ were defined in Chapters I and II. We shall

*In mathematical literature, they are sometimes called *Tikhonov stabilizers* (see, for example, [144–146]).

now show the connection (established in [74]) between these two concepts.

Suppose that

$$Az = u, \tag{2.3.6}$$

where z and u belong to Banach spaces F and U respectively and A is a linear operator from F into U, and suppose that the range $R(A)$ of the operator A is everywhere dense in U and that the inverse A^{-1} of A exists but it is not continuous. Let $\Omega[z]$ denote a continuous nonnegative convex functional defined on a linear manifold F_1 that is everywhere dense in F. Suppose that $\Omega[z]$ satisfies the following conditions:

a) $\Omega[0] = 0$;

b) for every fixed nonzero element z of F_1, the function $\varphi(\beta) = \Omega[\beta z]$ is a strictly increasing function of the variable β such that $\lim_{\beta \to +\infty} \varphi(\beta) = +\infty$;

c) for every $d \geqslant 0$, the set

$$F_d^1 \equiv \{z, z \in F_1, \Omega[z] \leqslant d\}$$

is compact.

Obviously, $\Omega[z]$ is a stabilizing functional for equation (2.3.6) and

$$F_1 = \bigcup_{d \geqslant 0} F_d^1.$$

We note that pth-order stabilizing functionals (see subsection 3) have the three properties listed.

Suppose that z_d is a quasisolution of equation (2.3.6) on a compact set F_d^1; that is, z_d is an element minimizing the functional $\| Az - u \|^2$ on the compact set F_d^1. It is shown in [79] that $\Omega[z_d] = d$. Consequently, for given $d > 0$, the problem of finding the quasisolution z_d on the compact set F_d^1 reduces to minimizing the functional

$$\rho_U^2(Az, u) = \| Az - u \|^2$$

under the condition $\Omega[z] = d$, that is, to the problem of finding the unconditional minimum of the functional

$$M^\alpha[z, u] = \rho_U^2(Az, u) + \alpha\Omega[z]$$

on the set F_d^1.

Suppose that an element z_α minimizes the functional $M^\alpha[z, u]$. By §3, the element z_α is a regularized solution of equation (2.3.6) on the set F_d^1. Thus, a quasisolution of equation (2.3.6) on the compact set F_d^1 is a regularized solution of that equation. When we substitute the element z_α into equation (2.3.6), we obtain the discrepancy $\rho_U(Az_\alpha, u) = \delta_\alpha$.

The three conditional parameters d, α, and δ_α are connected by two relationships:

$$\Omega[z_\alpha] = d, \quad \rho_U(Az_\alpha, u) = \delta_\alpha \tag{2.3.7}$$

If we know one of them, we can find the other two by using (2.3.7). With the regularization method, we are usually given the parameter δ_α, that is, the value of the error in the right-hand member of equation (2.3.6).

Since $F_{d_1}^1 \subset F_{d_2}^1$ for $d_1 < d_2$, a regularized solution of equation (2.3.6) on the set $F_{1,\delta} = F_1 \cap Q_\delta$ (see §2, Chapter II) belongs to the set $F_1 = \bigcup_{d \geqslant 0} F_d^1$.

Thus, the family of regularized approximate solutions of equation (2.3.6) consists of quasisolutions on the family of extended compact sets F_d^1.

5. A regularized solution of the equation $Az = u$ can also be constructed in the form of a series. Suppose that F and U are Hilbert spaces and that A is a completely continuous operator from F into U. Let F_1 denote a Hilbert subspace of the space F with a majorant norm such that, for every $d > 0$, the set of elements z of F_1 for which $\|z\| \leqslant d$ is compact in F. Then, we can take for the stabilizer the functional $\Omega[z] = \|z\|^2$. In this case

Euler's equation for the smoothing functional $M^\alpha[z, u]$ has the form

$$A^*Az + \alpha z = A^*u. \qquad (2.3.8)$$

Here, A^*A is a self-adjoint operator. Let $\{\varphi_n\}$ denote the complete system of its eigenfunctions and let $\{\lambda_n\}$ denote the corresponding eigenvalues.

As we know, A^*u can be represented in the form of a series

$$A^*u = \sum_{n=1}^{\infty} c_n \varphi_n.$$

If we seek a solution in the form

$$z = \sum_{n=1}^{\infty} b_n \varphi_n,$$

we obtain for its coefficients the formula

$$b_n = \frac{c_n}{\lambda_n + \alpha}.$$

The parameter α is determined from the discrepancy (cf. Theorem 3 of section 2 of Chapter I).

Variational methods of solving ill-posed problems are also examined in [23].

§4. Application of the regularization method to the approximate solution of integral equations of the first kind.

1. As was shown in §3, to find an approximate (regularized) solution of equation (2.0.1), it is sufficient to find an element $z_\alpha \in F$ minimizing the functional $M^\alpha[z, u]$. This last problem can be solved either by direct methods for minimizing the functional

(for example, by the method of steepest descent) or by solving the Euler equation corresponding to the functional $M^\alpha[z, u]$. This equation has the form

$$A^*Az + \alpha\Omega'[z] = A^*u,$$

where A^* is the adjoint of the operator A and $\Omega'[z]$ is the Fréchet derivative of the functional $\Omega[z]$.

2. Suppose that we are required to find a regularized solution of a Fredholm integral equation of the first kind on a finite interval $[a, b]$ (see [156])

$$\int_a^b K(x, s)\, z(s)\, ds = u(x), \tag{2.4.1}$$

where $c \leqslant x \leqslant d$ and $u(x) \in L_2[c, d]$.

Let us use a first-order stabilizer. Thus, we seek a regularized solution $z_\alpha(s)$ in the space W_2^1. It minimizes the functional

$$M^\alpha[z, u] = \int_c^d \left\{ \int_a^b K(x, s)\, z(s)\, ds - u(x) \right\}^2 dx + \\ + \alpha \int_a^b \left\{ q_0(s)\, z^2(s) + q_1(s) \left(\frac{dz}{ds} \right)^2 \right\} ds. \tag{2.4.2}$$

A condition for a minimum of this functional is vanishing of its first variation. This last condition is written in the form

$$\int_a^b \left(-\alpha \left\{ \frac{d}{ds}\left[q_1(s) \frac{dz}{ds} \right] - q_0(s)\, z(s) \right\} + \right. \\ \left. + \int_a^b \overline{K}(s, t)\, z(t)\, dt - b(s) \right) v(s)\, ds + \alpha q_1(s)\, z'(s)\, v(s) \Big|_a^b . \tag{2.4.3}$$

Here, $v(s)$ is an arbitrary variation of the function $z(s)$ such that both $z(s)$ and $z(s) + v(s)$ belong to the class of admissible functions,

$$\bar{K}(s, t) = \int_c^d K(\xi, s) K(\xi, t)\, d\xi,$$

and

$$b(s) = \int_c^d K(\xi, s)\, u(\xi)\, d\xi.$$

Condition (2.4.3) will be satisfied if

$$\int_a^b \bar{K}(s, t)\, z(t)\, dt - \alpha \left\{\frac{d}{ds}\left[q_1(s)\frac{dz}{ds}\right] - q_0(s)\, z\right\} = b(s) \tag{2.4.4}$$

and

$$q_1(s)\, z'(s)\, v(s)\,\Big|_a^b = 0. \tag{2.4.5}$$

Thus, if we know the values of the desired solution $z(s)$ of equation (2.4.1) at both ends of the interval $[a, b]$, then we can take as admissible functions in minimizing the functional (2.4.2) only functions $z(s)$ in W that assume the prescribed values at those end-points. In this case, the functions $v(s)$ must vanish at $s = a$ and $s = b$ and condition (2.4.5) will be satisfied.

In the case that we have described, the problem of finding a regularized solution $z_\alpha(s)$ thus reduces to finding a solution of the integrodifferential equation (2.4.4) satisfying the conditions

$$z(a) = \bar{z}_1, \quad z(b) = \bar{z}_2, \tag{2.4.6}$$

where $\bar{z}_1$ and $\bar{z}_2$ are known numbers.

If the values of the solution $z(s)$ that we are seeking are not known at the end-points $s = a$ and $s = b$, we can satisfy condition (2.4.5) by taking

$$z'(a) = z'(b) = 0. \tag{2.4.7}$$

In this case, we need to take as regularized solution of equation (2.0.1) a solution of equation (2.4.4) satisfying conditions (2.4.7). Other boundary conditions that the solution of equation (2.4.4) must satisfy are obviously possible. For example, there might be conditions of the form

$$z(a) = \overline{z_1}, \quad z'(b) = 0 \tag{2.4.8}$$

or

$$z'(a) = 0, \quad z(b) = \overline{z_2}. \tag{2.4.9}$$

Remark 1. When we do not know the values of the sought solution at the end-points of the interval, we set

$$z'(a) = 0, \quad z'(b) = 0.$$

But the solution of (2.4.1) that we are seeking may not satisfy these conditions. If, for example, we know that the derivative of the solution that we are seeking has the value g at $s = a$, then, by setting $z(s) = \tilde{z}(s) + g \cdot s$ in equation (2.4.1), we obtain an equation of the same form (with the same kernel) but with a different right-hand member for the function $\tilde{z}(s)$. The solution that we are seeking will satisfy the condition $\tilde{z}'(a) = 0$.

3. If in solving equation (2.4.1), we use pth-order stabilizers, then Euler's equation for the functional $M^{\alpha}[z, u]$ will have the form [157]

$$\int_a^b \overline{K}(s, t)\, z(t)\, dt + \alpha \sum_{r=0}^{p} (-1)^r \frac{d^r}{ds^r}\left[q_r(s)\frac{d^r z}{ds^r}\right] = b(s).$$

We can easily write boundary conditions that are possible for the solution sought for this equation. For example,

$$z(a) = z'(a) = \dots = z^{(p-1)}(a) = 0;$$
$$z(b) = z'(b) = \dots = z^{(p-1)}(b) = 0.$$

We shall not write other possible boundary conditions.

4. The problem (2.4.4), (2.4.6) or the problem (2.4.4), (2.4.7) and others can be solved numerically on a computer. In such a case, equation (2.4.4) is replaced with its finite-difference approximation on a given grid. If we take a uniform grid with step h, equation (2.4.4) is replaced with a system of finite-difference equations of the form

$$-\frac{\alpha}{h^2}\{q_{1,k-1}\cdot z_{k-1} + q_{1,k}\cdot z_{k+1} - (q_{1,k} + q_{1,k-1})\cdot z_k -$$
$$- h^2 q_{0,k}\cdot z_k\} + \sum_{r=0}^{n} \overline{K}_{k,r}\cdot z_r\cdot h = b_k; \quad k = 1,2,\dots,n-1. \tag{2.4.10}$$

Here, $q_{1,k}=q_1(s_k)$, $q_{0,k}=q_0(s_k)$, $z_k=z(s_k)$, $b_k=b(s_k)$, $s_k = kh + a$, $s_n = b$, and the $\overline{K}_{k,r}$ are the coefficients in the quadrature formula used to replace the integral in (2.4.4) with a finite sum. If the solution of equation (2.4.4) that we are seeking must satisfy the boundary conditions (2.4.6), then, in the system (2.4.10), we set

$$z_0 = \overline{z}_1, \quad z_n = \overline{z}_2.$$

On the other hand, if this solution satisfies conditions (2.4.7), the number k in the system (2.4.10) must assume the values

$$k = 0, 1, 2, \dots, n.$$

We then take

$$z_{-1} = z_0, \quad z_{n+1} = z_n.$$

Sometimes, it is more expedient to seek a solution on a nonuniform grid with mesh points s_k not necessarily equally spaced:

$$h_k = s_{k+1} - s_k, \quad h_k \neq h_{k+1}.$$

In these cases, equation (2.4.4) is approximated by a system of linear algebraic equations of the form

$$-\alpha \left\{ \frac{1}{h_k h_{k-1}} q_{1,k} z_{k+1} + \frac{1}{h_{k-1}^2} q_{1,k-1} z_{k-1} - \right.$$
$$\left. - \left(\frac{q_{1,k}}{h_k h_{k-1}} + \frac{q_{1,\,k-1}}{h_{k-1}^2} \right) z_k \right\} + \alpha q_{0,k} z_k + \sum_{r=0}^{n} \overline{K}_{k,r} \cdot z_r \cdot h_r = b_k. \tag{2.4.11}$$

The boundary conditions for the solution of this system are written just as in the case of a uniform grid (see also [48, 182, 183, 188, 210]).

Remark 2. An approximate solution of the integral equation (2.4.1) can also be constructed by replacing the integral with the corresponding approximating sum on the grid and replacing equation (2.4.1) with a system of linear algebraic equations. The system obtained is then solved by the regularization method described in Chapter III.

§5. Examples of application of the regularization method.

We next give some examples of the application of the regularization method to the solution of integral equations of the first kind.

Example 1. Consider the problem of numerical differentiation (see also [24, 51, 56, 57]). The nth derivative $z(t)$ of the function $u(t)$ is a solution of the integral equation

$$\int_0^t \frac{1}{(n-1)!} (t-\tau)^{n-1} z(\tau)\, d\tau = u(t). \tag{2.5.1}$$

If the right-hand member ($u = \tilde{u}(t)$) is given only approximately, we can speak only of an approximate expression for the derivative.

Suppose that $u(t) = \int_0^t \exp(-y^4)\,dy$. Then, $u'(t) = \exp(-t^4)$ and $u'''(t) = -12t^2 \exp(-t^4) + 16t^6 \exp(-t^4)$.

The graphs of the functions $u'(t)$ and $u'''(t)$ are shown by the solid curves in Figures 2 and 3.

Suppose that, instead of $u(t)$, we have $\tilde{u}(t_i) = u(t)(1 + \theta_i \epsilon)$, where the θ_i are random numbers in the interval $[-1, 1]$. Calculation of the first and third derivatives of the function $\tilde{u}(t)$ for one of the sequences of random numbers $\{\theta_i\}$ by the regularization method yields the results shown by the dashed curves in Figures 2 and 3. Here, with $u'(t)$ we took $h = 0.1$, $\epsilon = 0.1$, and $\alpha = 0.0055$; with $u'''(t)$, we took $h = 0.05$, $\epsilon = 0.01$, and $\alpha = 0.55 \cdot 10^{-6}$. The parameter α was determined from the discrepancy.

Let us look at two more examples, which are typical of problems associated with the processing of experimental observations.

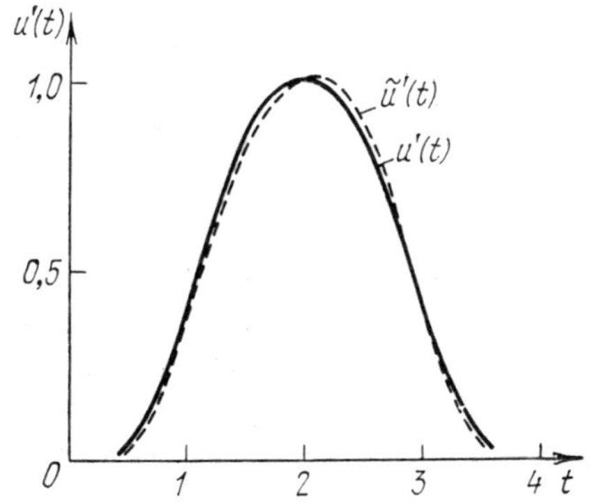

FIG. 2.

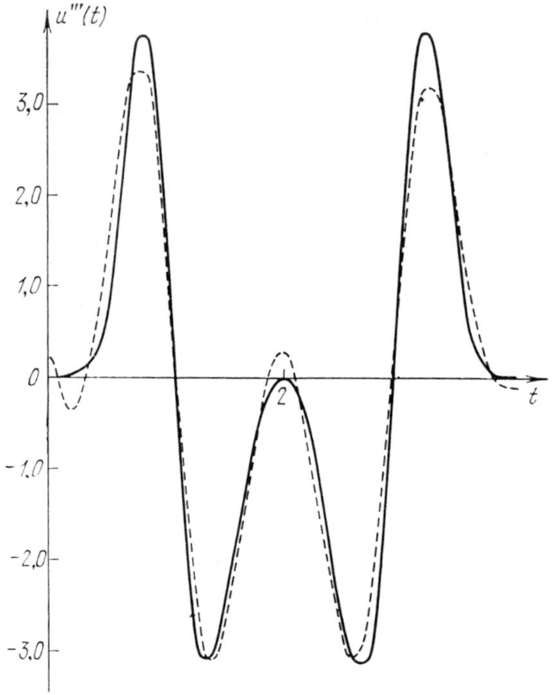

FIG. 3.

Example 2. Let us look at the problem of determining the spectral composition of radiation (electromagnetic radiation of the gamma ray type, X-rays, or corpuscular radiation) [162, 164, 175].

Suppose that the radiation that we are concerned with is nonhomogeneous and that the distribution of the density of the particles (photons) is characterized by a function $z(s)$, where s is either frequency or energy. When we test this radiation through a measuring device, we obtain an experimental spectrum $u(x)$ (where x may be frequency or energy). If the measuring device is linear, the connection between $z(s)$ and $u(x)$ is given by

$$Az \equiv \int_a^b K(x, s) z(s) ds = u(x), \qquad (2.5.2)$$

where a and b are the ends of the spectrum and $K(x, s)$ reflects a property of the device (assumed to be known) and is an experimental spectrum (with respect to x) if monochromatic radiation of frequency (or energy) s and of unit intensity falls on the device. This function can also be regarded as the response of the measuring device to the delta function $z = \delta(x - s)$.

The problem consists in determining the true spectrum of the radiation $z(s)$ from the experimental spectrum $u(x)$ and it reduces to solving equation (2.5.2) for $z(s)$.

Let us look at a mathematical model, taking $\bar{z}(s)$ close to the function $z_T(s)$ and a device function $K(x, s)$ close to the device function in the corresponding practical problems.

Solving the direct problem, let us calculate the experimental spectrum

$$\bar{u}(x) = \int_a^b K(x, s) \bar{z}(s) ds$$

on a grid with respect to x: $\{x_1, x_2, \ldots, x_n\}$. Simulating the process of bringing out random errors in the measurement of the experimental spectrum $u(x)$, we replace $\bar{u}(x_i)$ with $\tilde{u}(x_i)$ according to the formulas

$$\tilde{u}(x_i) = \bar{u}(x_i)\left(1 + \theta_i \sqrt{\frac{3(b-a)}{b^3 - a^3}}\,\sigma\right),$$

where the θ_i are random numbers in the interval $(-1, 1)$ with uniform distribution law. Obviously, the mean value of $\tilde{u}(x_i)$ is $\bar{u}(x_i)$ and the variance of $\tilde{u}(x_i)$ is σ^2. The mean square deviation

$$\|\tilde{u}(x) - \bar{u}(x)\| = \\ = \left\{\int_a^b [\tilde{u}(x) - \bar{u}(x)]^2 dx\right\}^{1/2} \approx \left[3\sigma^2 \frac{1}{n} \sum_i \theta_i^2\right]^{1/2} = \sigma$$

is a characteristic of the accuracy of the initial data.

Let us take for $\bar{z}(s)$ the function shown by the solid curve in Figure 4 and let us take $K(x, s) = \left(1 - \frac{s}{x}\right)\eta(x-s)$, where $\eta(x-s)$ is a unit function. Let us take $a = 0$ and $b = 11$. We calculate

$$\bar{u}(x) = \int_0^{11} K(x, s)\bar{z}(s)\,ds.$$

Then we solve the equation

$$\int_0^{11} K(x, s)\,z(s)\,ds = \bar{u}(x)$$

for $z(s)$.

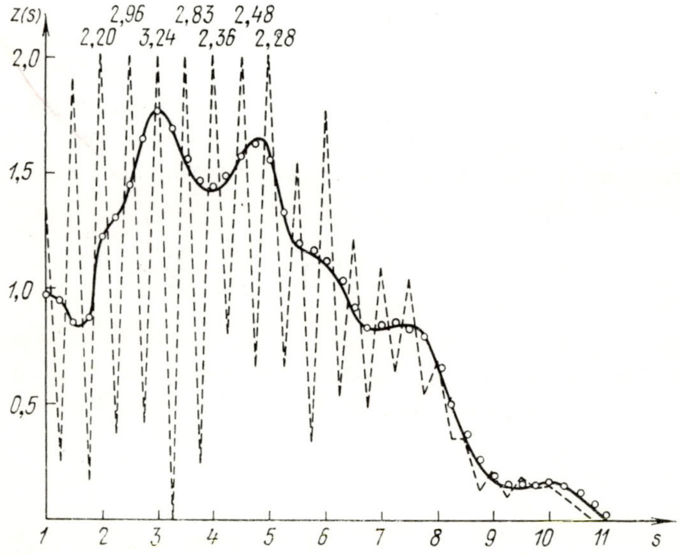

FIG. 4.

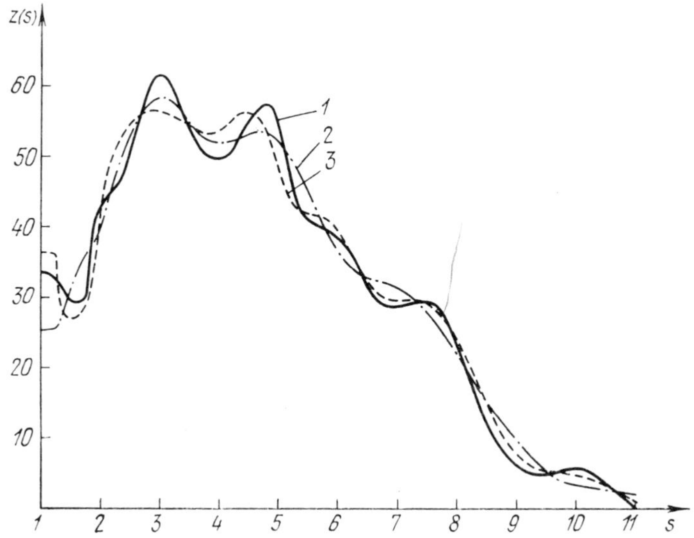

FIG. 5.

We replace the last equation with the system of linear algebraic equations

$$\sum_{j=1}^{n} K_{ij} z_j \Delta s_j = \bar{u}(x_i),$$

approximating the integral with a sum by means of Simpson's formula. The results are shown by the dashed broken line in Figure 4. The saw-toothed broken line has nothing in common with $\bar{z}(s)$. The circles in the same drawing show the values of z_j obtained by applying the regularization method. Here, the calculations were performed with machine accuracy. Figure 5 shows the results of applying the regularization method to the solution of the equation with right-hand member $\tilde{u}(x_i)$ with relative errors at the nodal points x_i equal to 5% and 10% of the measured value of $\bar{u}(x_i)$ (curves 2 and 3, curve 1 being the graph of $\bar{z}(x)$).

This example is a model for the problem of reproducing the true spectrum of a flow of fast neutrons caused by a polonium-berillium source. This spectrum is similar to the curve $\bar{z}(s)$. As recording device, we use a scintillation sensor [162].

For different levels of error in the right-hand member $u(x)$, we can, by solving the problem under consideration, estimate the corresponding errors in the function $z(s)$. In this way, we can predict the outcome of the experiment by determining $z(s)$ and thus plan the experiment.

This is a typical problem of mathematical planning of an experiment.

Example 3. Let us look at the problem of reproducing the shape of a time-dependent electrical impulse signal $z_T(t)$ applied at the input of a coaxial cable of length l from the output signal $u(t)$. The connection between $z_T(t)$ and $u(t)$ is given by

$$\int_0^t K(t-\tau)\, z_T(\tau)\, d\tau = u(t),$$

in which $K(t)$ is a known impulse function

$$K(t) = \eta(t) \frac{\mu l}{\sqrt{4\pi t^3}} \exp\left(-\frac{\mu^2 l^2}{4t}\right), \tag{2.5.3}$$

where μ is a constant characterizing the type of cable and $\eta(t)$ is a unit function.

For $z_T(t)$ we choose the function represented by the solid curve in Figure 6 and we take its convolution with the kernel (2.5.3), in which $\mu = 3.05 \cdot 10^{-4}$ and $l = 10^4$. We obtain the function $u(t)$ shown in Figure 7. At the nodal points of the grid $\{t_i\}$, we introduce into the values of $u(t_i) = \int_0^{t_i} K(t_i - \tau)\, z_T(\tau)\, d\tau$ disturbances according to the formulas $\tilde{u}(t_i) = u(t_i)(1 + \theta_i \epsilon)$, where the θ_i are random numbers in $[-1, 1]$. If we set $\epsilon = 0.01 \cdot p$, then

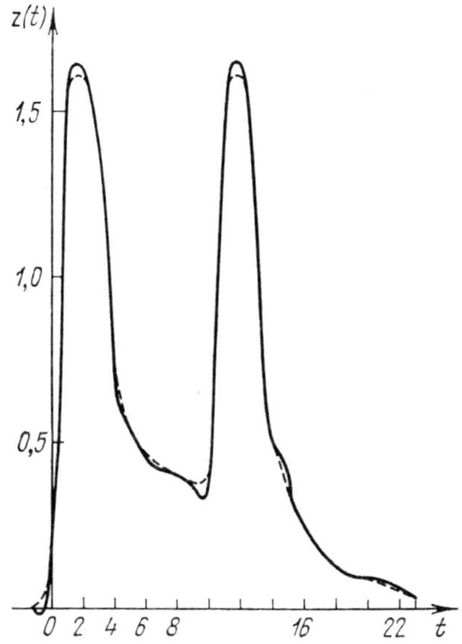

FIG. 6.

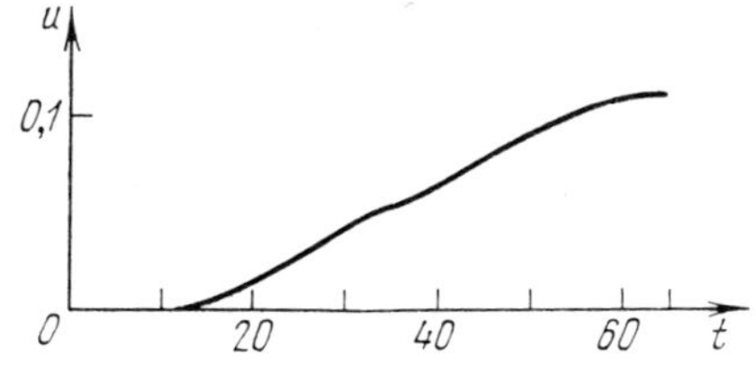

FIG. 7.

the relative error of $\widetilde{u}(t_i)$ in comparison with $u(t_i)$ does not exceed $p\%$. We took $p = 5$.

From the given values of $\widetilde{u}(t_i)$, we are required to find a solution of the equation

$$\int_0^t K(t-\tau)\, z(\tau)\, d\tau = \widetilde{u}(t)$$

that approximates $z_T(t)$.

This problem was solved by the regularization method on a grid of step $h = 0.4$ for $0 \leqslant \tau \leqslant 23$. For the stabilizing functional $\Omega[z]$ we took

$$\Omega[z] = \int_0^{23} \{(z')^2 + z^2\} dt.$$

For the boundary conditions we took $z(0) = z(23) = 0$. With $\alpha = 0.00156$, we obtained the results shown by the dashed curve in Figure 6. The parameter α was determined from the discrepancy.

Let us look at some inverse heat-flow problems. There are several kinds of these. We shall confine ourselves to two of them. The inverse heat-flow problem of solving the Cauchy problem for the heat-flow equation for negative time is well known. For example, we are required to solve the equation

$$a^2 u_{xx} = u_t$$

for $t < T$ if $u(x, T)$ is a known function $f(x)$. This problem is unstable under small changes in the function $f(x)$. Besides the above-described regularization method, another stable method of solving such problems is the method of quasireversibility (see [104]). A representative of the other kind of inverse heat-flow problem is the following, which is of great practical importance and leads to the integral equation examined in example 3:

Consider a homogeneous half-space $x > 0$ bounded by the plane $x = 0$. Suppose that the initial temperature is zero throughout this half-space and that the temperature on its boundary is a function $v(t)$ of time alone. In this case, the direct heat-flow problem consists in solving the equation $a^2 u_{xx} = u_t$ in the region ($x > 0$, $t > 0$) subject to the initial conditions $u(x, 0) = 0$, $u(0, t) = v(t)$. Its solution is

$$u(x, t) = \int_0^t \frac{x v(\tau)}{\sqrt{4\pi a^2 (t-\tau)^3}} \exp\left[\frac{-x^2}{4a^2 (t-\tau)}\right] d\tau.$$

The inverse heat-flow problem consists in finding the temperature $v(t)$ on the boundary $x = 0$ from the results of measuring the temperature $u(x_0, t)$ at a distance $x_0 > 0$ from the boundary. This problem reduces to solving the integral equation

$$\int_0^t \frac{x_0 v(\tau)}{\sqrt{4\pi a^2 (t-\tau)^3}} \exp\left[\frac{-x_0}{4a^2 (t-\tau)}\right] d\tau = u(x_0, t)$$

for $v(\tau)$, that is, to solving the equation examined in Example 3.

The heat-flow problem for negative time values are examined in [104, 203]. Similar questions are examined in [20, 34, 39, 55, 65].

§6. Determination of the regularization parameter.

1. The question of determination of the regularization parameter will be treated here only for regularizing operators $R_1(u, \alpha)$ obtained by the variational method (see §§ 1–3).

In the study of particular problems, it is usually difficult to actually find the regularization parameter α as a function $\alpha(\delta)$ (where δ is the error in the initial data) for which the operator $R_1(u, \alpha(\delta))$ is a regularizing operator. In many cases, we know a *number* δ characterizing the inaccuracy of the initial information. The problem consists in finding the corresponding value of the regularization parameter α out of all admissible values, that is, values that are equal to the value of one of the functions $\alpha = \alpha(\delta)$ for which the operator $R_1(u, \alpha(\delta))$ is a regularizing operator. The choice of the admissible value of the regularization parameter depends essentially on the information available regarding the approximate initial information. Various ways of finding such a value α are described in the literature. Below, we shall look at some of these.

First method. Suppose that the right-hand member u_δ of equation (2.0.1) is known with an error δ; that is, $\rho_U(u_\delta, u_T) \leqslant \delta$.

Then, under certain restrictions, the regularization parameter α can be determined from the discrepancy, that is, from

$$\rho_U(Az_\alpha, u_\delta) = \delta. \quad (2.6.1)$$

Let us denote $M^\alpha[z_\alpha, u]$, $\rho_U^2(Az_\alpha, u)$, and $\Omega[z_\alpha]$ by $m(\alpha)$, $\varphi(\alpha)$, and $\psi(\alpha)$ respectively. If the set of elements $F^\alpha \equiv \{z_\alpha\}$ for which $\inf M^\alpha[z, u]$ is attained consists of more than a single element, then $\varphi(\alpha)$ and $\psi(\alpha)$ will be multiple-valued functions. Let us look at certain properties of the functions $m(\alpha)$, $\varphi(\alpha)$, and $\psi(\alpha)$.

Lemma 1. *The functions* $m(\alpha)$ *and* $\varphi(\alpha)$ *are nondecreasing functions;* $\psi(\alpha)$ *is a nonincreasing function.*

Proof. Suppose that $\alpha_1 < \alpha_2$ and

$$\varphi_i = \rho_U^2(Az_{\alpha_i}, u),\ \psi_i = \Omega[z_{\alpha_i}],\ m_i = M^{\alpha_i}[z_{\alpha_i}, u] \quad (i = 1, 2),$$

where the z_{α_i} are arbitrary elements in the set F^{α_i}. We have

$$m_2 = \varphi_2 + \alpha_2\psi_2 \geqslant \varphi_2 + \alpha_1\psi_2 \geqslant \varphi_1 + \alpha_1\psi_1 = m_1, \quad (2.6.2)$$

from which the monotonicity of $m(\alpha)$ follows. Also,

$$\varphi_1 + \alpha_1\psi_1 \leqslant \varphi_2 + \alpha_1\psi_2 \text{ and } \varphi_2 + \alpha_2\psi_2 \leqslant \varphi_1 + \alpha_2\psi_1.$$

From this we get

$$(\alpha_1 - \alpha_2)\psi_1 \leqslant (\alpha_1 - \alpha_2)\psi_2.$$

Since $\alpha_1 < \alpha_2$, we have $\psi_1 \geqslant \psi_2$. It follows from this result and the second of inequalities (2.6.2) that $\varphi_2 \geqslant \varphi_1$. This completes the proof of the lemma.

Remark. By its definition, the function $m(\alpha)$ is single-valued. In contrast, if the set $F^\alpha \equiv \{z_\alpha\}$ consists of more than element, the functions $\varphi(\alpha)$ and $\psi(\alpha)$ may be multiple-valued since these may have different values for different elements z_α of F^α although the sum $\varphi(\alpha) + \alpha\psi(\alpha)$ is the single-valued function $m(\alpha)$. The monotonicity proven in the lemma applies to any choice of values of $\varphi(\alpha)$ and $\psi(\alpha)$.

Let us show that the functions $m(\alpha)$, $\varphi(\alpha)$, and $\psi(\alpha)$ are lower- and upper-semicontinuous. In the following lemmas, $\{\alpha_n\}$ will denote a sequence of positive numbers that converges to a positive number α_0 and $\{z_{\alpha_n}\}$ will denote the sequence of the corresponding elements z_{α_n} of the sets F^{α_n}.

Lemma 2. *If the sequence* $\{z_{\alpha_n}\}$ *converges, then*

$$\lim_{n\to\infty} z_{\alpha_n} = \bar{z} \in F^{\alpha_0}. \tag{2.6.3}$$

Proof. Obviously,

$$\lim_{n\to\infty} M^{\alpha_n}[z_{\alpha_n}, u] = M^{\alpha_0}[\bar{z}, u]$$

since the terms of the functional $M^\alpha[z, u]$ are continuous with respect to z and α. Let us suppose that the element $\bar{z}$ does not belong to the set F^{α_0}, that is, that it does not minimize the functional $M^{\alpha_0}[z, u]$. Then, there exists $z^1_{\alpha_0} \in F^{\alpha_0}$ such that

$$M^{\alpha_0}[z^1_{\alpha_0}, u] = M^{\alpha_0}[\bar{z}, u] - \beta, \text{ where } \beta > 0.$$

This leads to a contradiction since

$$\lim_{n\to\infty} M^{\alpha_n}[z_{\alpha_0}, u] = M^{\alpha_0}[z_{\alpha_0}, u] = M^{\alpha_0}[\bar{z}, u] - \beta,$$

so that, beginning with some $n(\beta)$, we have for all $n \geqslant n(\beta)$

$$M^{\alpha_n}[z_{\alpha_0}, u] < M^{\alpha_0}[\bar{z}, u] - \frac{\beta}{2}.$$

On the other hand,

$$M^{\alpha_n}[z_{\alpha_n}, u] > M^{\alpha_0}[\bar{z}, u] - \frac{\beta}{2}.$$

Consequently,

$$M^{\alpha_n}[z_{\alpha_n}, u] > M^{\alpha_n}[z_{\alpha_0}, u],$$

which contradicts the definition of the element z_{α_n}. This completes the proof of the lemma.

Lemma 3. *The functions* $m(\alpha)$, $\varphi(\alpha)$, *and* $\psi(\alpha)$ *are lower- and upper-semicontinuous at every* $\alpha > 0$.

We shall confine ourselves to proving that $\varphi(\alpha)$ is lower-semicontinuous since the proofs for the other two functions and for upper-semicontinuity are the same.

Proof. Suppose that $\{\alpha_n\}$ is an increasing sequence of positive numbers that converges to a positive number α_0. Corresponding to it is the sequence $\{F^{\alpha_n}\}$ of sets of elements z_{α_n} that minimize the functional $M^{\alpha_n}[z, u]$. Let $\{z_{\alpha_n}\}$ denote an arbitrary sequence of elements z_{α_n} of F^{α_n}. Beginning with some n, the terms of this sequence belong to the compact set

$$\{z; \Omega[z] \leqslant \Omega[z_{\alpha_0-\epsilon}], \epsilon > 0\}.$$

Consequently, it has a convergent subsequence. Without loss of generality, we may assume that $\{z_{\alpha_n}\}$ is that subsequence. We write $\lim_{n\to\infty} z_{\alpha_n} = \bar{z}$. It follows from Lemma 2 that $\bar{z} \in F^{\alpha_0}$. Then, the sequence of values of

$$\rho_U(Az_{\alpha_n}, u)$$

converges to

$$\rho_U(Az_{\alpha_0}, u).$$

By Lemma 1, the sequence $\{\varphi(\alpha_n)\}$ is nondecreasing and thus converges to a number $\bar{\varphi}$, which is the greatest lower bound of the set of values of $\{\varphi(\alpha_0)\}$. If this were not the case, there would exist a value $\tilde{\varphi}(\alpha_0)$ less than $\bar{\varphi}$. Then, for sufficiently large n, there would exist elements z_{α_n} of F^{α_n} for which $\varphi(\alpha_n)$ is also less than $\bar{\varphi}$, which violates the monotonicity of the function $\varphi(\alpha)$. It follows that, for an arbitrary subsequence of the original sequence $\{z_{\alpha_n}\}$, the corresponding subsequence $\{\varphi(\alpha_n)\}$ converges to $\bar{\varphi}$. This means that the entire sequence $\{\varphi(\alpha_n)\}$ converges to $\bar{\varphi}$. This proves the lower-semicontinuity.

Remembering that

$$m(\alpha) = M^{\alpha}[z_{\alpha}, u]$$

is, by its definition, a single-valued function of α, we immediately obtain the

Corollary. *The function $m(\alpha)$ is a continuous nondecreasing function of α.*

We note that, if the set AF is everywhere dense in U, then $m(\alpha) \to 0$ as $\alpha \to 0$ (and $m(0) = 0$). This follows from the fact that, for every $\epsilon > 0$, there exists an element z^1 and an $\alpha = \alpha(\epsilon)$ such that

$$M^{\alpha}[z^1, u] = \rho_U^2(Az^1, u) + \alpha\Omega[z^1] < \epsilon$$

for $\alpha = \alpha(\epsilon)$. For this, we need to choose z^1 so that

$$\rho_U^2(Az^1, u) < \frac{\epsilon}{2} \text{ and } \alpha(\epsilon) < \frac{\epsilon}{2\Omega[z^1]}.$$

The possibility of doing this follows from the fact that AF is everywhere dense in U.

We note also that $\varphi(0) = 0$. This is true because

$$\varphi(\alpha) + \alpha\psi(\alpha) = m(\alpha) \to 0 \quad \text{as} \quad \alpha \to 0.$$

From these lemmas, we immediately get the

Theorem. *If $\varphi(\alpha)$ is a single-valued function, then, for every positive number*

$$\delta < \rho_U(Az_0, u), \textit{ where } z_0 \in \{z; \Omega[z] = \inf_{Y \in F_1} \Omega[Y]\},$$

there exists an $\alpha(\delta)$ such that

$$\rho_U(Az_{\alpha(\delta)}, u) = \delta.$$

Remark. The function $\varphi(\alpha)$ is single-valued, for example, if the element z_α is unique (see p. 63).

In computational practice, one way of determining α from the discrepancy is as follows:

Let δ denote the error in the right-hand member u_δ of equation (2.0.1). We take a finite section of the monotonic sequence $\alpha_0, \alpha_1, \alpha_2, \ldots, \alpha_n$. For example, the α's may constitute a geometric progression: $\alpha_k = \alpha_0 q^k$ for $k = 0, 1, 2, \ldots, n$, where $q > 0$. For every value of α_k, one finds an element (function) z_{α_k} minimizing the functional

$$M^{\alpha_k}[z, u_\delta]$$

and one calculates the discrepancy $\rho_U(Az_{\alpha_k}, u_\delta)$. One chooses for α a number α_{k_0} for which, with the required accuracy,

$$\rho_U(Az_{\alpha_{k_0}}, u_\delta) = \delta.$$

An approximate solution of the equation $\varphi(\alpha) = \rho_U(Az_\alpha, u_\delta) = \delta$ for α can also be found by Newton's method. We need only note that the function $\varphi_1(\gamma) = \varphi(1/\gamma)$ is a decreasing convex function [115, 120]. Therefore, for an arbitrary initial approximation $\gamma_0 = 1/\alpha_0 > 0$, Newton's method converges. The derivative of $\varphi_1(\gamma)$, which we need when using this method, will be expressed in terms of the derivative $y_\alpha = dz_\alpha/d\alpha$ of the regularized solution z_α with respect to α. The element z_α is a solution of Euler's equation for the functional $M^\alpha[z, u_\delta]$:

$$(A^*A + \alpha B)z = A^*u_\delta, \tag{2.6.4}$$

where $B = \Omega'[z]$.

When we differentiate the identity $(A^*A + \alpha B)z_\alpha \equiv A^*u_\delta$ with respect to α, we find that $y_\alpha = dz_\alpha/d\alpha$ is a solution of the equation

$$(A^*A + \alpha B)y = -\alpha B z_\alpha, \tag{2.6.5}$$

which differs from Euler's equation (2.6.4) only in the right-hand member.

The problem [120] of finding, out of all elements of the set F_1 that satisfy the condition $\Omega[z] \leqslant R^2$, an element that provides $\inf \rho^{U_2}(Az, u_\delta)$ also leads to the problem of minimizing the functional $M^\alpha[z, u_\delta]$. Therefore, if we know the number R, we can determine the regularization parameter α from the condition $\Omega[z_\alpha] = R^2$. This is the second method of determining α.

In a manner analogous to the proof of the solvability of the equation $\rho_U(Az_\alpha, u) = \delta$, we can show that the equation $\Omega[z_\alpha] = R^2$ is solvable. In computational practice, this value of α can be found approximately either by a sorting from a given set of values $\alpha_1, \alpha_2, \ldots, \alpha_n$ or by Newton's method [115, 116], which (as is shown in the two articles mentioned) converges for an arbitrary initial approximation $\alpha_0 > 0$.

We note that the derivative $y_\alpha = dz_\alpha/d\alpha$, which is involved here, is a solution of equation (2.6.5).

The *third method* consists in finding a *quasioptimal* value of the regularization parameter $\alpha = \alpha_{qo}$.

Let us assume the space F to be equipped with a norm. By definition (see [67]),

$$\alpha_{qo} = \inf_{\alpha} \sup_{\rho_U(u_\delta, u_T) \leqslant \delta} \left\| \alpha \frac{dz_\alpha}{d\alpha} \right\|_F ,$$

where the supremum is over all the right-hand members u_δ of equation (2.0.1) that satisfy the inequality $\rho_U(u_\delta, u_T) \leqslant \delta$. To find α_{qo} approximately, we need to find the regularized solutions z_α corresponding to a large number of possible right-hand members u_δ. Frequently, we have only one specific right-hand member u_δ. In such cases, we can find only

$$\inf_{\alpha} \left\| \alpha \frac{dz_\alpha}{d\alpha} \right\|_F .$$

The efficiency of this method of determining the parameter α for certain classes of problems has been shown with representative examples (see [67, 160, 165]). We note that the function $y_\alpha = \alpha dz_\alpha / d\alpha$ is a solution of the equation

$$(A^*A + \alpha B)\, y = A^*Az_\alpha - A^*u_\delta,$$

which differs from Euler's equation for the functional $M^\alpha[z, u_\delta]$ only in the right-hand member.

The *fourth method* consists in taking, as a suitable value of the regularization parameter, a value $\alpha = \alpha_0$ that *maximizes* the ratio

$$v(\alpha) = \frac{\rho_U\left(A\left(\alpha \frac{dz_\alpha}{d\alpha}\right),\ Az_\alpha - u_\delta\right)}{\rho_U(Az_\alpha, u_\delta)}$$

The efficiency of this method of choosing α has been shown [67] for sample problems with the aid of computational experiments.

Depending on the supplementary information that we may have regarding the right-hand member, there are yet other methods of choosing the parameter α. For example, in Chapter IV, §2 we shall give a method of finding a (c, f)-optimal value of the regularization parameter, which makes use of similar information. The efficiency of the different methods of determining the parameter α for different classes of problems is established by performing a computational experiment. Analogous questions are examined in [28, 49, 50].

CHAPTER III

SOLUTION OF SINGULAR AND ILL-CONDITIONED SYSTEMS OF LINEAR ALGEBRAIC EQUATIONS

1. We know the difficulties involved in solving so-called ill-conditioned* systems of linear algebraic equations. Large changes (beyond what is acceptable) in the solution may result from small changes in the right-hand members of such systems.

Consider the system of equations

$$Az = u, \tag{3.0.1}$$

where $A = \{a_{ij}\}$ is a matrix with elements a_{ij}, $z = \{z_j\}$ is the unknown vector with coordinates z_j, and $u = \{u_i\}$ is a known vector with coordinates u_i. In these definitions, $i, j = 1, 2, \ldots, n$.

The system (3.0.1) is said to be **singular** if the determinant of A is zero. In this case, the matrix A has zero eigenvalues.

If we consider systems with fixed norming of the elements of the matrix A, then the determinant $\det A$ is close to zero for ill-conditioned systems of this kind.

---

*It should be pointed out that this term does not have a fully established definition.

If the calculations are only approximate, it is impossible in some cases to determine whether a given system of equations is singular or ill-conditioned. Thus, ill-conditioned and singular systems can be indistinguishable within the framework of a given accuracy. Obviously, such a situation may arise when the matrix A has eigenvalues sufficiently close to zero.

In practical problems, we often know only approximately the right-hand member u of the system and the elements of the matrix A, that is, the coefficients in the system (3.0.1). In such cases, we are dealing not with the system (3.0.1) but with some other system $\tilde{A}z = \tilde{u}$ such that $\|\tilde{A} - A\| \leqslant \delta$ and $\|\tilde{u} - u\| \leqslant \delta$, where the particular norm chosen usually depends on the nature of the problem. Since we have the matrix $\tilde{A}$ rather than the matrix A, we cannot make a definite judgment as to the singularity or nonsingularity of the system (3.0.1).

In such cases, all we know about the exact system $Az = u$ whose solution we need to find is that $\|\tilde{A} - A\|$ and $\|\tilde{u} - u\|$ are each no greater than δ. But there are infinitely many systems with such initial data (A, u), and, within the framework of the error level known to us, they are indistinguishable. Since we have the approximate system $\tilde{A}z = \tilde{u}$ instead of the exact system (3.0.1), we can speak only of finding an approximate solution. But the approximate system $\tilde{A}z = \tilde{u}$ may not be solvable. The question then arises as to what we should understand by an approximate solution of the system (3.0.1).

Among the "possible exact systems" there may be singular systems. If a singular system has any solution at all, it has infinitely many. In such a case, when we speak of an approximation of the solution, which solution do we mean?

Thus, we often have to consider a whole class of systems of equations that are indistinguishable from each other (on the basis of a given error level) that may include both singular and unsolvable systems. The methods of constructing approximate solutions of systems of this class must be generally applicable. These solutions must be stable under small changes in the initial data in (3.0.1).

The construction of such methods is based on the idea of

"selection" expounded in Chapter II in the discussion of the regularization method.

2. Thus, suppose that the system (3.0.1) is singular and that the vector u constituting the right-hand member satisfies the conditions of solvability of the system. The solution of such a system is not unique. Let F_A denote the set of its solutions. We then pose the problem of finding a normal solution of the system. Following [166, 167], we define the **normal solution** of the system (3.0.1) for the vector z^1 as the solution z^0 for which

$$\|z^0 - z^1\| = \inf_{z \in F_A} \|z - z^1\|,$$

where z^1 is a fixed element (vector) determined by the formulation of the problem and $\|z\|$, the norm of the vector z, is defined by

$$\|z\| = \left\{\sum_{j=1}^{n} z_j^2\right\}^{1/2}.$$

In what follows, we shall assume for simplicity that $z^1 = 0$. Obviously, the normal solution is unique.

Remark 1. The normal solution z^0 of the system (3.0.1) can be defined as the solution minimizing a given positive-definite quadratic form in the coordinates of the vector $z - z^1$. All the results to be given below remain valid with this definition.

Remark 2. Suppose that the rank of the matrix A of a singular system (3.0.1) is $r < n$. Let $\bar{z}_{r+1}, \bar{z}_{r+2}, \ldots, \bar{z}_n$ constitute a basis of the vector space N_A consisting of elements z for which $Az = 0$; that is, $N_A = \{z;\ Az = 0\}$. The solution $\bar{z}^0$ of the system (3.0.1) that satisfies the $n - r$ orthogonality conditions

$$(\bar{z}^0 - z^1, \bar{z}_s) = 0, \quad s = r + 1, r + 2, \ldots, n, \qquad (3.0.2)$$

is unique and, as one can easily see, it coincides with the normal solution.

Remark 3. In linear algebra, a vector $\tilde{z}$ minimizing the discrepancy $\|Az - u\|^2$ is called a **pseudosolution** of the system (3.0.1). A normal pseudosolution is defined analogously to our definition of a normal solution. The method described below for constructing approximate normal solutions of the system (3.0.1) can also be used to find approximate normal pseudosolutions (see [121]).

3. One can easily see that the problem of finding the normal solution of the system (3.0.1) is ill-posed. Suppose that A is a symmetric matrix. If it is nonsingular, an orthogonal transformation

$$z = Vz^*, \quad u = Vu^*$$

will put it in diagonal form, and the resulting system will have the form

$$\lambda_i \overset{*}{z_i} = \overset{*}{u_i} \qquad (i = 1, 2, \ldots, n),$$

where the λ_i are the eigenvalues of the matrix A.

If the symmetric matrix A is singular of rank r, then $n - r$ of its eigenvalues will be equal to zero. Suppose that

$$\lambda_i \neq 0 \quad \text{for} \quad i = 1, 2, \ldots, r$$

and

$$\lambda_i = 0 \quad \text{for} \quad i = r + 1, r + 2, \ldots, n.$$

Suppose that the "initial data" (A and u) of the system are given with an error; that is, instead of A and u we are given δ-approximations $\tilde{A}$ and $\tilde{u}$ of them:

$$\|\tilde{A} - A\| \leqslant \delta, \ \|\tilde{u} - u\| \leqslant \delta.$$

Here

$$\|A\| = \left\{\sum_{i,j} a_{ij}^2\right\}^{1/2}, \; \|u\| = \left\{\sum_i u_i^2\right\}^{1/2}. \qquad (3.0.3)$$

Suppose that $\tilde{\lambda}_i$ are the eigenvalues of the matrix $\tilde{A}$. We know that they depend continuously on A in the norm (3.0.3). Consequently, the eigenvalues $\tilde{\lambda}_{r+1}, \tilde{\lambda}_{r+2}, \ldots, \tilde{\lambda}_n$ can be made arbitrarily small for sufficiently small δ.

If they are not equal to zero, then

$$\tilde{z}_i^* = \frac{1}{\tilde{\lambda}_i} \tilde{u}_i^* .$$

Thus, there exist disturbances of the system within the limits of any sufficiently small tolerance δ for which certain $\bar{z}_i{}^*$ will assume arbitrary prenamed values. This means that the problem of finding the normal solution of the system (3.0.1) is unstable.

Below, we shall describe a method, developed in [166, 167], for finding the normal solution of the system (3.0.1) that is stable under small disturbances (in the norm (3.0.3)) of the right-hand member u and of the matrix A. It is based on the regularization method.

§1. The regularization method of finding the normal solution.

1. Suppose that, instead of the exact singular system $Az = \bar{u}$, we have the system with approximate right-hand member

$$Az = \tilde{u}, \qquad (3.1.1)$$

where $\|\tilde{u} - \bar{u}\| \leqslant \delta$ and the vector $\tilde{u}$ may fail to satisfy the solvability condition.

It is natural to seek an approximate normal solution of the system (3.1.1) among the vectors z such that $\|Az - \tilde{u}\| \leqslant \delta$. By the definition of normal solution, this solution will minimize the functional $\Omega[z] = \|z - z^1\|^2$.

Thus, the problem reduces to minimizing the functional $\|z-z^1\|^2$ on the set of vectors satisfying the inequality $\|Az-\tilde{u}\|\leqslant\delta$. Since the functional $\Omega[z]=\|z-z^1\|^2$ is obviously a stabilizing and quasimonotonic functional (see Chapter II, §2), the latter problem is equivalent to the problem of minimizing that functional on the set of vectors z satisfying the condition $\|Az-\tilde{u}\|=\delta$. It reduces to finding the vector z^α minimizing the smoothing functional

$$M^\alpha[z, \tilde{u}, A] = \|Az - \tilde{u}\|^2 + \alpha\|z - z^1\|^2, \quad \alpha > 0.$$

The value of the parameter α is then determined from the condition $\|Az^\alpha - \tilde{u}\|=\delta$, that is, from the discrepancy.

Obviously, there exists only one such vector z^α. It can be determined from the system of linear equations

$$\alpha z_k^\alpha + \sum_{j=1}^{n} \bar{a}_{k,j} z_j^\alpha = \tilde{b}_k, \qquad k = 1, 2, \ldots, n$$

where

$$\bar{a}_{k,j} = \sum_{i=1}^{n} a_{ik} a_{ij} \text{ and } \tilde{b}_k = \sum_{i=1}^{n} a_{ik} \tilde{u}_i,$$

or with the aid of some other algorithm for minimizing the functional (form)

$$M^\alpha[z, \tilde{u}, A].$$

The vector z^α can be regarded as the result of applying to $\tilde{u}$ some operator $z^\alpha = R(\tilde{u}, \alpha)$ depending on the parameter α. Since the conditions for applicability of the regularization method hold in the present case, it follows from what was said in Chapter II. §2 that the operator $R(\tilde{u}, \alpha)$ is a regularizing operator. Hence, the vector $z^\alpha = R(\tilde{u}, \alpha)$ can be used as approximate normal solution of the system (3.1.1).

With an eye to broadening the field of applicability of the regularization method later on, let us again prove some theorems on the asymptotic behavior of $z^{\alpha(\delta)}$ as $\delta \to 0$.

We denote by U_A the vector subspace of the vectors z (for $z \in R^n$):

$$U_A \equiv \{u;\, u = Az,\, z \in R^n\}.$$

Let $\tilde{v}_A$ denote the projection of the vector $\tilde{u}$ onto U_A. Obviously,

$$\|\tilde{u} - \tilde{v}_A\| \leqslant \|\tilde{u} - Az\|.$$

Theorem 1. *As $\alpha \to 0$, the vectors z^α minimizing the functional $M^\alpha[z, \tilde{u}, A]$ converge to the normal solution $\hat{z}_0$ of the system $Az = \tilde{v}_A$.*

Proof. Let $\hat{F}_A$ denote the set of all solutions of the system $Az = \tilde{v}_A$. As we know, the vector $v = \tilde{u} - \tilde{v}_A$ is orthogonal to the subspace U_A. Consequently,

$$\|Az - \tilde{u}\|^2 = \|Az - \tilde{v}_A\|^2 + \|\tilde{u} - \tilde{v}_A\|^2$$

and

$$M^\alpha[z, \tilde{u}, A] = \|Az - \tilde{v}_A\|^2 + \|\tilde{u} - \tilde{v}_A\|^2 + \alpha\Omega[z] =$$
$$= M^\alpha[z, \tilde{v}_A, A] + \|\tilde{u} - \tilde{v}_A\|^2. \quad (3.1.2)$$

Since $\tilde{v}_A \in U_A$, the system

$$Az = \tilde{v}_A \quad (3.1.3)$$

is solvable. The vector z^α minimizing the functional $M^\alpha[z, \tilde{u}, A]$ also minimizes the functional $M^\alpha[z, \tilde{v}_A, A]$ since, by (3.1.2), these functionals differ by $\|\tilde{u} - \tilde{v}_A\|^2$, which is independent of z.

Since $A\hat{z}^0 = \tilde{v}_A$, we have

$$\alpha\Omega[z^\alpha] \leqslant M^\alpha[z^\alpha, \tilde{v}_A, A] \leqslant M^\alpha[\hat{z}^0, \tilde{v}_A, A] = \alpha\Omega[\hat{z}^0]$$

or

$$\Omega[z^{\alpha}] \leqslant \Omega[\hat{z}^0] = d. \tag{3.1.4}$$

Since R^n is an n-dimensional Euclidean space and $z \in R^n$, the set of vectors z such that

$$\Omega[z] = \|z\|^2 \leqslant d$$

is compact for arbitrary $d > 0$.

Thus, z^{α} belongs to a compact set. From this set it is possible to construct a sequence $\{z^{\alpha_k}\}$ that converges to some element z_1^0 as $\alpha_k \to 0$. The vector z_1^0 minimizes the functional $\|Az - \tilde{v}_A\|^2$. Since the minimum of this functional is 0, we have

$$Az_1^0 = \tilde{v}_A;$$

that is, the vector z_1^0 is a solution of equation (3.1.3). It follows from (3.1.4) that

$$\Omega[z_1^0] \leqslant \Omega[\hat{z}^0]$$

or

$$\|z_1^0\| < \|\hat{z}^0\|. \tag{3.1.5}$$

Since the normal solution $\hat{z}^0$ of equation (3.1.3) minimizes the norm $\|\hat{z}^0\|$, it follows from (3.1.5) that $z_1^0 = \hat{z}^0$. This completes the proof of the theorem.

2. Let us now look at the case in which both the right-hand member of the equation and the matrix A are inexactly given; that is, let us look at an equation of the form

$$\tilde{A}z = \tilde{u}, \tag{3.1.6}$$

where

$$\|\tilde{u} - u\| \leqslant \delta, \quad \|\tilde{A} - A\| \leqslant \delta.$$

At the end of §2 of Chapter II, we proved the applicability of the variational principle to the construction of regularized approximate solutions of such problems. This principle reduces to minimization of the corresponding smoothing functional $M^{\alpha}[z, \tilde{u}, \tilde{A}]$. The regularization parameter can be determined with the use of this method from the generalized discrepancy [47]. Below, we shall give a theorem on the asymptotic behavior of the regularized solution $z^{\alpha(\delta)}$ as $\delta \to 0$. The possibility of determining the regularization parameter by other methods follows.

Thus, instead of looking at this problem, we shall follow [166, 167] and look at the problem of minimizing the functional (form)

$$M^{\alpha}[z, \tilde{u}, \tilde{A}] = \|\tilde{A}z - \tilde{u}\|^2 + \alpha\Omega[z],$$

where

$$\Omega[z] = \|z\|^2.$$

There exists a unique element $\tilde{z}^{\alpha}$ minimizing the functional $M^{\alpha}[z, \tilde{u}, \tilde{A}]$.

Let us suppose that the vector u satisfies the conditions for solvability and that z^0 is the normal solution of the equation $Az = u$.

Theorem 2. *Suppose that $\tilde{A}$ and $\tilde{u}$ are δ-approximations of the matrix A and the vector u. Let $\beta(\delta)$ and $\alpha_0(\delta)$ denote continuous positive functions that approach 0 monotonically as $\delta \to 0$ and that satisfy the inequality*

$$\frac{\delta^2}{\beta(\delta)} \leqslant \alpha_0(\delta).$$

Then, for every $\epsilon > 0$, there exists a $\delta_0(\epsilon, \|z^0\|) > 0$ such that the inequality

$$\|\tilde{z}^{\alpha} - z^0\| \leqslant \epsilon$$

holds for every α satisfying the inequalities

$$\frac{\delta^2}{\beta(\delta)} \leqslant \alpha \leqslant \alpha_0(\delta), \tag{3.1.7}$$

where $0 < \delta \leqslant \delta_0$.

Proof. Define $U_{\tilde{A}} \equiv \{u; u = \tilde{A}z, z \in R^n\}$ and let $\tilde{v}_{\tilde{A}}$ denote the projection of the vector $\tilde{u}$ onto the subspace $U_{\tilde{A}}$. Then, the vector $\tilde{v} = \tilde{u} - \tilde{v}_{\tilde{A}}$ is orthogonal to $U_{\tilde{A}}$. We know that

$$\|\tilde{u} - \tilde{v}_{\tilde{A}}\| \leqslant \|\tilde{u} - \tilde{A}z\|$$

and

$$M^\alpha[z, \tilde{u}, \tilde{A}] = \|\tilde{u} - \tilde{v}_{\tilde{A}}\|^2 + M^\alpha[z, \tilde{v}_{\tilde{A}}, \tilde{A}].$$

Consequently, the vector $\tilde{z}^\alpha$ minimizing the functional

$$M^\alpha[z, \tilde{u}, \tilde{A}]$$

also minimizes the functional

$$M^\alpha[z, \tilde{v}_{\tilde{A}}, \tilde{A}].$$

Obviously,

$$M^\alpha[\tilde{z}^\alpha, \tilde{v}_{\tilde{A}}, \tilde{A}] \leqslant M^\alpha[z^0, \tilde{v}_{\tilde{A}}, \tilde{A}] = \|\tilde{A}z^0 - \tilde{v}_{\tilde{A}}\|^2 + \alpha\Omega[z^0].$$

One can easily see that

$$\|\tilde{A}z^0 - \tilde{v}_{\tilde{A}}\| \leqslant \|\tilde{A}z^0 - Az^0\| + \|Az^0 - \tilde{v}_{\tilde{A}}\| \leqslant C\delta, \tag{3.1.8}$$

where

$$C = 2(1 + \|z^0\|).$$

Specifically,

$$\|\tilde{A}z^0 - Az^0\| \leqslant \|\tilde{A} - A\| \cdot \|z^0\| \leqslant \delta\|z^0\|. \tag{3.1.9}$$

Then, using inequality (3.1.7) and the equation $Az^0 = u$, we find

$$\|Az^0 - v_{\tilde{A}}\| = \|u - v_{\tilde{A}}\| \leqslant \|u - \tilde{u}\| + \|\tilde{u} - v_{\tilde{A}}\| \leqslant \|u - \tilde{u}\| + \|\tilde{u} - \tilde{A}z^0\| \leqslant \|u - \tilde{u}\| + \|\tilde{u} - u\| + \|u - \tilde{A}z^0\| = 2\|\tilde{u} - u\| + \|Az^0 - \tilde{A}z^0\| \leqslant 2\|\tilde{u} - u\| + \|\tilde{A} - A\| \, \|z^0\| \leqslant \delta\,(2 + \|z^0\|).$$

Thus,

$$\|Az^0 - v_{\tilde{A}}\| \leqslant \delta\,(2 + \|z^0\|). \tag{3.1.10}$$

Inequality (3.1.8) then follows from (3.1.9) and (3.1.10). Furthermore,

$$\alpha\Omega[\tilde{z}^\alpha] \leqslant M^\alpha[\tilde{z}^\alpha, v_{\tilde{A}}, \tilde{A}] \leqslant M^\alpha[z^0, v_{\tilde{A}}, \tilde{A}] = \|\tilde{A}z^0 - v_{\tilde{A}}\|^2 + \alpha\Omega[z^0].$$

Using inequality (3.1.8), we obtain

$$\alpha\Omega[\tilde{z}^\alpha] \leqslant M^\alpha[\tilde{z}^\alpha, v_{\tilde{A}}, \tilde{A}] \leqslant \leqslant C^2\delta^2 + \alpha\Omega[z^0] = \alpha\left[\frac{C^2\delta^2}{\alpha} + \|z^0\|^2\right], \tag{3.1.11}$$

or

$$\|\tilde{z}^\alpha\|^2 \leqslant \frac{C^2\delta^2}{\alpha} + \|z^0\|^2. \tag{3.1.12}$$

For arbitrary α satisfying conditions (3.1.7), we find from (3.1.12)

$$\|\tilde{z}^\alpha\| \leqslant \|z^0\| + \epsilon_1(\delta), \tag{3.1.13}$$

where $\epsilon_1(\delta) \to 0$ as $\delta \to 0$.

Let $\{\delta_n\}$, $\{\alpha_n\}$, $\{\tilde{A}_n\}$, and $\{\tilde{u}_n\}$ denote sequences, respectively, of positive numbers, positive numbers, matrices, and right-hand members of equation (3.1.1) such that

$$\delta_n \to 0 \text{ as } n \to \infty,$$

$$\|\tilde{A}_n - A\| \leqslant \delta_n, \ \|\tilde{u}_n - u\| \leqslant \delta_n,$$

and

$$\frac{\delta_n^2}{\beta(\delta_n)} \leqslant \alpha_n \leqslant \alpha_0(\delta_n). \tag{3.1.14}$$

Corresponding to these sequences is a sequence of vectors $\tilde{z}^{\alpha_n}$ minimizing respectively the functionals

$$M^{\alpha_n}[z, \tilde{u}_n, \tilde{A}_n].$$

Since the numbers α_n satisfy inequalities (3.1.14), they also satisfy inequalities (3.1.13); that is, for every n,

$$\|\tilde{z}^{\alpha_n}\| \leqslant \|z^0\| + \epsilon_1(\delta_n). \tag{3.1.15}$$

This means that the sequence $\{\tilde{z}^{\alpha_n}\}$ is bounded (with respect to the norm) and hence compact. Therefore, it has a convergent subsequence. Without changing the notation, we obtain

$$\lim_{n \to \infty} \tilde{z}^{\alpha_n} = z_1^0.$$

For vectors in this subsequence, we have

$$\begin{aligned} \|A\tilde{z}^{\alpha_n} - Az^0\| &\leqslant \\ &\leqslant \|A\tilde{z}^{\alpha_n} - \tilde{A}_n \tilde{z}^{\alpha_n}\| + \|\tilde{A}_n \tilde{z}^{\alpha_n} - v_{\tilde{A}}\| + \|v_{\tilde{A}} - Az^0\| \leqslant \\ &\leqslant \delta_n \|\tilde{z}^{\alpha_n}\| + \sqrt{M^{\alpha_n}[\tilde{z}^{\alpha_n}, v_{\tilde{A}_n}, \tilde{A}_n]} + \delta_n(2 + \|z^0\|) \leqslant \\ &\leqslant \delta_n\{2 + 2\|z^0\| + \epsilon_1(\delta_n)\} + \sqrt{C^2\delta_n^2 + \alpha_0(\delta_n)\|z^0\|^2} = b_n. \end{aligned}$$

Here, we used the following inequalities: (3.1.10) for an estimate of the norm $\|v_{\tilde{A}_n} - Az^0\|$, (3.1.15) for an estimate of $\|\tilde{z}^{\alpha_n}\|$, and (3.1.11) for an estimate of $M^{\alpha_n}[\tilde{z}^{\alpha_n}, v_{\tilde{A}_n}, \tilde{A}_n]$. Obviously, $b_n \to 0$ as $n \to \infty$. Taking the limit in

$$\| \tilde{A}\tilde{z}^{\alpha_n} - Az^0 \| = \| \tilde{A}\tilde{z}^{\alpha_n} - u \| \leqslant b_n,$$

we obtain

$$Az_1^0 = u.$$

Thus, z_1^0 is a solution of the equation $Az = u$. It coincides with the normal solution z^0 by virtue of the normal solution's minimality and the inequality

$$\| z_1^0 \| \leqslant \| z^0 \|,$$

which is obtained by taking the limit in inequality (3.1.15). This completes the proof of the theorem.

This theorem justifies our taking $\tilde{z}^\alpha$ as approximate normal solution of the system.

Remark 1. Theorem 2 holds for a broad class of equations of the form $\tilde{A}z = \tilde{u}$ with disturbed initial data $(\tilde{A}, \tilde{u})$. Its proof remains virtually unchanged.

Remark 2. If the matrix $\tilde{A}$ is ill-conditioned and its δ-neighborhood contains a singular matrix A, the conditions are then equivalent to the conditions of Theorem 2 and, in this case, it is natural to solve the system by the regularization method as described above.

Questions of stable methods of solving systems of algebraic equations have also been examined in [29, 30, 62-64, 118, 121].

§ 2. Supplementary remarks.

Remark 1. Since the exposition of §1 is not essentially connected with the finite dimensionality of the spaces F and U to which the elements z and u belong, that exposition remains valid for arbitrary continuous linear operators Az (the proofs are repeated word for word) if U is a Hilbert space and F belongs to a normed space in which F is s-compactly embedded [161]. This enables us to use the regularization method for Fredholm integral equations of the second kind (see [2, 3]).

Remark 2. The method described is also suitable for solving ill-posed linear programming problems (see Chapter VIII) in which one seeks a solution of a system that satisfies supplementary constraints (the solution must belong to a closed convex set).

Remark 3. The stabilizer $\Omega[z]$ can also be taken in the form

$$\Omega[z] = \sum_{i=1}^{n} \rho_i (z_i - z_{1i})^2,$$

where $\rho_i > 0$ and $z_1 = (z_{1,1}, z_{1,2}, \ldots, z_{1,n})$.

This change of stabilizer does not change either the formulations of the theorems of §1 or their proofs.

CHAPTER IV

APPROXIMATE SOLUTIONS OF INTEGRAL EQUATIONS OF THE FIRST KIND OF THE CONVOLUTION TYPE

Among integral equations of the first kind, one often encounters equations of the convolution type $K(t)*z(t) = u(t)$, for example, an equation of the form

$$\int_{-\infty}^{\infty} K(t-\tau) z(\tau)\, d\tau = u(t). \tag{4.0.1}$$

Obviously, the regularization method can be applied to the construction of approximate solutions of. equations of the convolution type. For this it is sufficient to indicate methods of constructing regularizing operators. In Chapter II, we examined a variational method for constructing such operators. In the present chapter, we shall, following [11, 14, 15], use integral transformations for equations of the convolution type to construct a broad family of regularizing operators that are easily realized on a computer. For one subclass of this class, we shall indicate the connection with the regularizing operators obtained by the variational method.

We shall look in detail at the construction of a regularizing operator for equation (4.0.1) with the use of the Fourier transformation. However, all the results obtained (and their proofs) will be analogous for equations of the convolution type that require use of the Laplace transformation, Mellin's transformation, and others. Therefore, we shall not give the reasoning or the calculations here for the corresponding equations.

It is shown in §1 that the error $v(t)$ in the right-hand member $u(t)$ is additive; that is, $u(t) = u_T(t) + v(t)$.

The class of regularizing operators mentioned above can be constructed under very weak requirements on the right-hand member $u(t)$ of the equation and on the error $v(t)$ (u and v belong to L_2) and without allowance for the random nature of the function $v(t)$.

In examining the deviation (see §2) of the regularized solution from the exact one, we assume in addition that the error $v(t)$ (interference, noise) is the sample function of a stationary random process that is uncorrelated with the solution $z_T(t)$ that we are seeking. This deviation is estimated in the probabilistic metric $\sup_t \overline{[z_\alpha(t) - z_T(t)]^2}$, where the vinculum denotes the mathematical expectation. Below, we shall examine regularized solutions obtained with the aid of very simple pth-order stabilizers.

According to the nature of the asymptotic behavior (as $\omega \to \infty$) of the Fourier transform of the kernel of the equation, we can distinguish four classes of integral equations of the convolution type encountered in practice for which, under supplementary assumptions regarding the asymptotic behavior (as $\omega \to \infty$) of the spectral density of the error $v(t)$, asymptotic estimates with respect to α (as $\alpha \to 0$) for the deviation of the regularized solution from the exact one [9–11, 13] are given both for the part due to the influence of the regularization method and for the part due to the error in the right-hand member.

§1. Classes of regularizing operators for equations of the convolution type.

1. Consider an equation of the convolution type

$$z(t) * K(t) = u(t), \qquad (4.1.1)$$

in which $K(t)$ and $u(t)$ are given functions and $z(t)$ is the unknown function. Here, z belongs to a metric space F and u belongs to a metric space U.

Let us suppose that, for $u(t) = u_T(t)$, this equation has a unique solution $z_T(t)$ belonging to F, that is,

$$z_T(t) * K(t) \equiv u_T(t).$$

The problem consists in finding the function $z_T(t)$. If the right-hand member is known with an error, that is, if, instead of $u_T(t)$, we have a function $u(t)$ such that

$$\rho_U(u_T, u) \leqslant \delta,$$

then, instead of finding z_T, we can pose only the problem of finding an approximate solution. As approximate solution, let us take the regularized solution

$$z_\alpha(t) = R(u, \alpha),$$

where $R(u, \alpha)$ is a regularizing operator.

2. Let us look at the broad class of regularizing operators obtained with the aid of classical integral transformations. We shall describe this class.

Suppose that F is a set of functions belonging to L_1 and that U is a set of functions belonging to L_2. Let A denote a continuous linear operator with domain of definition D_A containing F. Consider the equation

$$Az = u, \quad u \in U. \qquad (4.1.2)$$

Let us suppose that this equation has a unique solution on F. Let $\mathcal{J}$ denote a linear integral transformation, for example, the Fourier transformation (or it could be the Laplace or Mellin transformation). Application of $\mathcal{J}$ to (4.1.2) yields

$$\mathcal{F}[Az] = \mathcal{F}[u] = u(\omega). \tag{4.1.3}$$

Suppose that the operator A is such that we can use (4.1.3) to determine $\mathcal{F}[z] = z(\omega)$ in the form

$$z(\omega) = \psi(u(\omega), \omega).$$

If Az is the convolution $z(t)*K(t)$ of the function $z(t)$ with some given function $K(t)$ (the kernel) and if the convolution formula

$$\mathcal{F}[z*K] = \mathcal{F}[z] \cdot \mathcal{F}[K]$$

holds for the transformation $\mathcal{F}$, then

$$\psi(u(\omega), \omega) = \frac{u(\omega)}{K(\omega)}, \tag{4.1.4}$$

where $K(\omega)$ is the $\mathcal{F}$-transform of the function $K(t)$. Formula (4.1.4) is valid as applied, for example, to convolutions of the form

a) $$z(t) * K(t) = \int_{-\infty}^{\infty} K(t-\tau) z(\tau) \, d\tau$$

if we use the Fourier transformation or of the form

b) $$z(t) * K(t) = \int_{0}^{t} K(t-\tau) z(\tau) \, d\tau$$

(here $K(t) \equiv z(t) \equiv 0$ for $t < 0$) if we use the one-sided Laplace transformation or of the form

c) $$z(t) * K(t) = \int_{0}^{\infty} K\left(\frac{t}{\tau}\right) z(\tau) \frac{d\tau}{\tau}$$

if we use the Mellin transformation.

3. For definiteness, let us look at an equation of the form

$$Az \equiv \int_{-\infty}^{\infty} K(t-\tau) z(\tau)\, d\tau = u(t) \qquad (4.1.5)$$

and let us apply the Fourier transformation. Here,

$$u(t) \in L_2(-\infty, \infty),$$

$$K(t) \in \mathscr{K} \subset L_1(-\infty, \infty),$$

and

$$z(t) \in F \subset L_1(-\infty, \infty).$$

If the right-hand member of equation (4.1.5) is known only approximately, that is, if $u(t) = u_T(t) + v(t)$, where $v(t)$ is an interference (noise), then

$$z(\omega) = \frac{u(\omega)}{K(\omega)} = \frac{u_T(\omega)}{K(\omega)} + \frac{v(\omega)}{K(\omega)}.$$

Since $u_T(\omega) = K(\omega) z_T(\omega)$, we have

$$z(\omega) = z_T(\omega) + \frac{v(\omega)}{K(\omega)}.$$

This formula gives us the Fourier transform of the exact solution of equation (4.1.5) with approximate right-hand member $u(t)$. As approximate solution of equation (4.1.5) with approximate right-hand member $u(t)$, it would seem natural to take the function obtained with the aid of the inverse Fourier transformation, that is, the function

$$z(t) = \frac{1}{2\pi} \int_{-\infty}^{\infty} z(\omega) \exp(-i\omega t)\, d\omega =$$

$$= \frac{1}{2\pi} \int_{-\infty}^{\infty} z_T(\omega) \exp(-i\omega t)\, d\omega + \frac{1}{2\pi} \int_{-\infty}^{\infty} \frac{v(\omega)}{K(\omega)} \exp(-i\omega t)\, d\omega =$$

$$= z_T(t) + \frac{1}{2\pi} \int_{-\infty}^{\infty} \frac{v(\omega)}{K(\omega)} \exp(-i\omega t)\, d\omega.$$

However, this function may not exist since the last integral may diverge. Specifically, by the properties of the Fourier transformation, the functions $K(\omega)$ and $v(\omega)$ approach 0 as $\omega \to \infty$, but this approach to zero is not coordinated since the function $v(t)$ (hence $v(\omega)$) is usually of a random nature. Therefore, the ratio

$$v(\omega)/K(\omega)$$

may, as a result of the influence of high frequencies ω of the random function $v(\omega)$, fail to have an inverse Fourier transform. But even if this ratio has an inverse Fourier transform $w(t)$, the deviation of the function $w(t)$ from zero (in the C- or L_2-metric) can be arbitrarily large.

Thus, we cannot take the exact solution of equation (4.1.5) as approximate solution of that equation with approximate right-hand member. Such a solution may not exist and, even if it does, it may not be stable under small deviations in the right-hand member $u(t)$. The reason for the instability of such an algorithm for constructing the "solutions" is the influence of high frequencies ω of the Fourier transform $v(\omega)$ of the interference $v(t)$. Therefore, if we wish to construct approximate solutions of equation (4.1.5) that are stable under small deviations in the right-hand member $u(t)$ with the aid of the inverse Fourier transformation, we need to "suppress" the influence of the high frequencies ω, multiplying, for example, the function

$$u(\omega)/K(\omega)$$

by a suitable multiplier $f(\omega, \alpha)$ (depending on the parameter α in accordance with §1 of Chapter II) [11, 14, 15].

4. Returning to equation (4.1.2), let us look at an operator of the form

$$R_f(u, \alpha) = \mathcal{F}^{-1}[\psi(u(\omega), \omega) \cdot f(\omega, \alpha)],$$

where $\mathcal{F}^{-1}$ is the inverse of the transformation $\mathcal{F}$ and $f(\omega, \alpha)$ is a given function defined for all nonnegative values of the parameter α and arbitrary ω that occur in forming the operator $\mathcal{F}^{-1}$. Under suitable conditions on the function $f(\omega, \alpha)$, the operator $R_f(u, \alpha)$ is a regularizing operator for equation (4.1.2).

For definiteness, we shall examine in what follows an equation of convolution type of the form (4.1.5) and shall take for $\mathcal{F}$ the Fourier transformation. In this case,

$$R_f(u, \alpha) \equiv \frac{1}{2\pi} \int_{-\infty}^{\infty} \frac{f(\omega, \alpha)}{K(\omega)} u(\omega) \exp(-i\omega t)\, d\omega. \qquad (4.1.6)$$

Suppose that the function $f(\omega, \alpha)$ satisfies the following conditions:

1_f) $f(\omega, \alpha)$ is defined on the region $(\alpha \geqslant 0, -\infty < \omega < \infty)$;

2_f) $0 \leqslant f(\omega, \alpha) \leqslant 1$ for all $\alpha \geqslant 0$ and ω;

3_f) $f(\omega, 0) \equiv 1$;

4_f) for every $\alpha > 0$, the function $f(\omega, \alpha)$ is an even function with respect to ω and it belongs to $L_2(-\infty, \infty)$;

5_f) for every $\alpha > 0$, we have $f(\omega, \alpha) \to 0$ as $\omega \to \pm\infty$;

6_f) $f(\omega, \alpha) \to 1$ nondecreasingly as $\alpha \to 0$ and this convergence is uniform on every interval $|\omega| \leqslant \omega_1$;

7_f) $f(\omega, \alpha)/K(\omega) \in L_2(-\infty, \infty)$ for every $\alpha > 0$;

8_f) for every $\omega \neq 0$, we have $f(\omega, \alpha) \to 0$ as $\alpha \to \infty$, and this convergence is uniform on every interval of the form $[\omega_1, \omega_2]$, where $0 < \omega_1 < \omega_2$.

Thus, a function $f(\omega, \alpha)$ satisfying these eight conditions defines a one-parameter operator $R_f(u, \alpha)$ of the form (4.1.6).

5. If the deviation of the right-hand member of equation (4.1.5) is estimated in the metric of $L_2(-\infty, \infty)$ but the deviation of the solution $z(t)$ is estimated in the C-metric and if we assume that $z(\omega) \in L_1(-\infty, \infty)$, then we have the

Theorem. *If the function $f(\omega, \alpha)$ satisfies conditions 1_f–8_f, the corresponding operator $R_f(u, \alpha)$ of the form* (4.1.6) *is a regularizing operator for equation* (4.1.5) *that is continuous with respect to u.*

Proof. Let us show that the operator $R_f(u, \alpha)$ is continuous with respect to u. We estimate the difference

$$\Delta R_f = R_f(u_1, \alpha) - R_f(u_2, \alpha) = \\ = \frac{1}{2\pi} \int_{-\infty}^{\infty} \frac{f(\omega, \alpha)}{K(\omega)} [u_1(\omega) - u_2(\omega)] \exp(-i\omega t)\, d\omega.$$

We have

$$|\Delta R_f| \leqslant \frac{1}{2\pi} \int_{-\infty}^{\infty} \left| \frac{f(\omega, \alpha)}{K(\omega)} \right| | u_1(\omega) - u_2(\omega) |\, d\omega.$$

Applying the Cauchy-Bunyakovskiy inequality, we obtain

$$|\Delta R_f| \leqslant \frac{1}{2\pi} \left\{ \int_{-\infty}^{\infty} \left| \frac{f}{K} \right|^2 d\omega \right\}^{1/2} \left\{ \int_{-\infty}^{\infty} | u_1(\omega) - u_2(\omega) |^2 d\omega \right\}^{1/2}.$$

By property 7_f, the integral

$$\varphi(\alpha) = \int_{-\infty}^{\infty} \left| \frac{f(\omega, \alpha)}{K(\omega)} \right|^2 d\omega$$

exists and hence the value of $\varphi(\alpha)$ is finite. The deviation of $u_1(t)$ from $u_2(t)$ in the L_2-metric, that is, $\rho_{L_2}(u_1, u_2)$ is, by

Plancherel's theorem,* equal to

$$\frac{1}{2\pi}\left\{\int_{-\infty}^{\infty}|u_1(\omega)-u_2(\omega)|^2 d\omega\right\}^{1/2}.$$

If $\rho_{L_2}(u_1, u_2)\leqslant\delta$, then $|\Delta R_f|\leqslant\delta\sqrt{\varphi(\alpha)}$. The continuity of the operator $R_f(u, \alpha)$ follows.

If the operator $R(u, \alpha)$ is continuous with respect to u and if $\lim_{\alpha\to 0} R(Az, \alpha)=z$ for every $z\in F$, then this operator is, by the theorem of Chapter II, §1, a regularizing operator. Therefore, to prove the theorem, it will be sufficient to show that, for every function $z(t)$ in F such that $z(\omega)\in L_1$, we have

$$\lim_{\alpha\to 0} R_f(Az, \alpha) = z(t),$$

where

$$Az \equiv \int_{-\infty}^{\infty} K(t-\tau)z(\tau)\,d\tau.$$

Since $u(\omega) = K(\omega)z(\omega)$, we have

$$|\Delta z| = |R_f(u, \alpha) - z(t)| = \\ = \frac{1}{2\pi}\left|\int_{-\infty}^{\infty} z(\omega)\{f(\omega, \alpha)-1\}\exp(-i\omega t)\,d\omega\right|.$$

Consequently,

---

*For every function $u(t)$ in L_2,

$$\int_{-\infty}^{\infty}|u(\omega)|^2 d\omega=2\pi\int_{-\infty}^{\infty}|u(t)|^2 dt.$$

$$|\Delta z| \leqslant \frac{1}{2\pi} \int_{-\infty}^{\infty} |z(\omega)|\,|f(\omega, \alpha) - 1|\,d\omega =$$

$$= \frac{1}{2\pi} \int_{-\infty}^{-\omega_1} |z(\omega)|\,|f(\omega, \alpha) - 1|\,d\omega + \frac{1}{2\pi} \int_{\omega_1}^{\infty} |z(\omega)|\,|f(\omega, \alpha) - 1|\,d\omega +$$

$$+ \frac{1}{2\pi} \int_{-\omega_1}^{\omega_1} |z(\omega)|\,|f(\omega, \alpha) - 1|\,d\omega.$$

Since $0 \leqslant f(\omega, \alpha) \leqslant 1$, it follows that

$$|\Delta z| \leqslant \frac{1}{2\pi} \int_{-\infty}^{-\omega_1} |z(\omega)|\,d\omega + \frac{1}{2\pi} \int_{\omega_1}^{\infty} |z(\omega)|\,d\omega +$$

$$+ \frac{1}{2\pi} \int_{-\omega_1}^{\omega_1} |z(\omega)|\,|f(\omega, \alpha) - 1|\,d\omega.$$

Since $z(\omega) \in L_1(-\infty, \infty)$, for every $\epsilon > 0$ there exists an $\omega_1(\epsilon) > 0$ such that the inequality

$$\frac{1}{2\pi} \int_{-\infty}^{-\omega_1} |z(\omega)|\,d\omega + \frac{1}{2\pi} \int_{\omega_1}^{\infty} |z(\omega)|\,d\omega \leqslant \frac{\epsilon}{2}$$

holds for every $\omega_1 \geqslant \omega_1(\epsilon)$. By property 6_f of the function $f(\omega, \alpha)$, there exists an $\alpha_0(\epsilon)$ such that the inequality

$$\frac{1}{2\pi} \int_{-\omega_1(\epsilon)}^{\omega_1(\epsilon)} |z(\omega)|\,|f(\omega, \alpha) - 1|\,d\omega \leqslant \frac{\epsilon}{2}.$$

holds for every $\alpha \leqslant \alpha_0(\epsilon)$. Thus, for given $\epsilon > 0$, there exists an $\alpha_0(\epsilon)$ such that, for $\alpha \leqslant \alpha_0(\epsilon)$,

$$|\Delta z| = |R_f(Az, \alpha) - z(t)| \leqslant \epsilon.$$

This means that

$$\lim_{\alpha \to 0} R_f(Az, \alpha) = z(t).$$

This completes the proof of the theorem.

Thus, the function $z_\alpha(t) = R_f(u, \alpha)$ obtained with the aid of the operator (4.1.6) is a regularized solution of equation (4.1.5). If the function of two variables $f(\omega, \alpha)$ of the operator (4.1.6) possesses properties 1_f–8_f, we shall refer to it as a **stabilizing factor**.

Remark 1. In the article [75] on solution of the Cauchy problem for Laplace's equation in a strip, a stabilizing factor of the form

$$f(\omega, \alpha) = e^{-\alpha^2\omega^2}$$

is used.

Remark 2. If we take

$$f(\omega, \alpha) = \begin{cases} 1, & |\omega| \leqslant 1/h; \\ 0, & |\omega| > 1/h; \end{cases} \quad (\alpha = h),$$

where h is the step of the grid on which we are seeking a solution of equation (4.1.5), we obtain the familiar method of constructing an approximate solution of that equation.

6. Suppose that $M(\omega)$ is a given even function with the following properties:

a) It is piecewise-continuous on every finite interval;

b) it is nonnegative and $M(\omega) > 0$ for $\omega \neq 0$;

c) $M(\omega) \geqslant C > 0$ for sufficiently large $|\omega|$;

d) for every $\alpha > 0$, the ratio $K(-\omega)/[L(\omega) + \alpha M(\omega)]$ belongs to $L_2(-\infty, \infty)$, where $L(\omega) = K(\omega)K(-\omega) = |K\omega|^2$.

If we set

$$f(\omega, \alpha) = \frac{L(\omega)}{L(\omega) + \alpha M(\omega)},$$

we obtain classes of regularizing operators for equation (4.1.5). Each such class is determined by the function $M(\omega)$.

7. For a particular function $M(\omega)$, the regularization parameter α can be found from the resulting discrepancy. If the discrepancy in the right-hand member $u(t)$ is estimated in the L_2-metric, the square of the discrepancy for the regularized solution $z_\alpha(t)$ is calculated from the formula

$$\rho^2_{L_2}(Az_\alpha, u) = \int_{-\infty}^{\infty} [Az_\alpha - u(t)]^2\, dt =$$

$$= \frac{1}{2\pi} \int_{-\infty}^{\infty} |K(\omega) z_\alpha(\omega) - u(\omega)|^2 d\omega = \Phi(\alpha).$$

Since

$$z_\alpha(\omega) = \frac{K(-\omega)\, u(\omega)}{L(\omega) + \alpha M(\omega)},$$

we have

$$\Phi(\alpha) = \frac{1}{2\pi} \int_{-\infty}^{\infty} \frac{\alpha^2 M^2(\omega)\, |u(\omega)|^2 d\omega}{\{L(\omega) + \alpha M(\omega)\}^2}.$$

Obviously, $\Phi(0) = 0$ and

$$\Phi(\alpha) \leqslant \frac{1}{2\pi} \int_{-\infty}^{\infty} |u(\omega)|^2 d\omega = \|u(t)\|^2_{L_2}.$$

Also, $\Phi(\alpha)$ approaches $\|u(t)\|^2_{L_2}$ as $\alpha \to \infty$. In addition,

$$\Phi'(\alpha) = \frac{1}{\pi}\int_{-\infty}^{\infty} \frac{\alpha L(\omega) M^2(\omega) \mid u(\omega) \mid^2 d\omega}{\{L(\omega) + \alpha M(\omega)\}^3} > 0.$$

Thus, the discrepancy resulting from the regularized solution is a strictly increasing function of the variable α and it ranges from 0 to $\|u(t)\|_{L_2}$. Consequently, if the error δ in the right-hand member u_δ of equation (4.1.5) is less than $\|u_\delta(t)\|_{L_2}$, then there exists a unique solution $\bar{\alpha}$ for which $\Phi(\bar{\alpha}) = \delta^2$. On the other hand, if $\delta \geqslant \|u_\delta(t)\|_{L_2}$, the equation $\Phi(\alpha) = \delta^2$ does not have a solution.

8. Obviously, $\Phi(\alpha)$ depends on the choice of the function $M(\omega)$; that is,

$$\Phi(\alpha) = \Phi_M(\alpha).$$

For small values of α, the basic contribution to $\Phi_M(\alpha)$ is made by large frequencies. Therefore, if $M_1(\omega) \geqslant M_2(\omega)$ for sufficiently large ω, then, for sufficiently small α,

$$\Phi_{M_1}(\alpha) \geqslant \Phi_{M_2}(\alpha).$$

In particular, if we take $M_1(\omega) = \omega^{2p_1}$ and $M_2(\omega) = \omega^{2p_2}$, then, for $p_1 > p_2$,

$$\Phi_{p_1}(\alpha) > \Phi_{p_2}(\alpha).$$

The solution $\bar{\alpha}$ of the equation $\Phi_p(\alpha) = \delta^2$ obviously depends on p: $\bar{\alpha} = \bar{\alpha}(p)$. Also, for $p_1 > p_2$, we have $\bar{\alpha}(p_1) < \bar{\alpha}(p_2)$. Thus, with increase in the order of regularization p, the value of $\bar{\alpha}(p)$ decreases.

What was said in §6 of Chapter II applies here to the determining of $\bar{\alpha}$ from the discrepancy.

9. One easily sees that the regularized solution of equation (4.1.5) defined by

$$z_\alpha(t) = \frac{1}{2\pi}\int_{-\infty}^{\infty} \frac{K(-\omega)\, u(\omega) \exp(-i\omega t)}{L(\omega) + \alpha M(\omega)} d\omega \qquad (4.1.7)$$

minimizes the functional

$$M^{\alpha}[z, u] = \int_{-\infty}^{\infty} (Az - u)^2\, dt + \alpha\Omega[z]$$

with stabilizing functional of the form

$$\Omega[z] = \int_{-\infty}^{\infty} M(\omega)\,|\,z(\omega)\,|^2 d\omega. \qquad (4.1.8)$$

Here,

$$Az \equiv \int_{-\infty}^{\infty} K(t - \tau)\, z(\tau)\, d\tau.$$

If we take

$$M(\omega) = \sum_{n=0}^{p} q_n \omega^{2n},$$

where the q_n are given nonnegative constants and, in particular, $q_p > 0$, we obtain from formula (4.1.8) pth-order stabilizers. If we take $M(\omega) = \omega^{2r}$, where r is any positive number, we obtain stabilizers of the form

$$\Omega[z] = B \int_{-\infty}^{\infty} |\,z^{(r)}(t)\,|^2\, dt,$$

where

$$z^{(r)}(t) = \frac{\Gamma(m+1)}{\Gamma(m+1-r)} \int_{-\infty}^{t} (t - \tau)^{m-r} \frac{d^{m+1}z}{d\tau^{m+1}}\, d\tau \,\cdot$$

(The function $z^{(r)}(t)$ can be regarded as the derivative of nonintegral order r.) Here, m is an integer at least equal to the integral part of r (that is, $m \geqslant [r]$) and B is a positive number.

We may take for $M(\omega)$ a function with arbitrary rate of growth as $\omega \to \infty$. Obviously, proceeding as we did with the examples given, we can obtain other stabilizers $\Omega[z]$ on the basis of formula (4.1.8).

10. The reason for considering different families of regularizing operators is that we might choose for each particular problem (or class of problems) the best operator. This might be, for example, an operator minimizing the discrepancy (in the sense defined above) between the regularized solution $z_\alpha(t)$ and the exact solution $z_T(t)$ that we are seeking, or it might be an operator that is more convenient for machine realization.

Let $\Phi(z_1, z_2)$ denote a given nonnegative functional defined on a set of the form $S \times S$, where S is a set of functions to which both the regularized solutions $z_\alpha(t)$ of equation (4.1.5) and the exact solution $z_T(t)$ belong. (For example, $\Phi(z_1, z_2)$ might be the norm of the difference between z_2 and z_1).

Whatever the functional $\Phi(z_1, z_2)$ might be, let us look at the following problems:

Problem 1. For a fixed stabilizing factor $f(\omega, \alpha)$, find a value α_0 of the regularization parameter α such that

$$\Phi(z_{\alpha_0}, z_T) = \inf_{\alpha} \Phi(z_\alpha, z_T).$$

Problem 2. Let $\mathcal{F}_f = \{f(\omega, \alpha)\}$ denote a given family of stabilizing factors. From this family, find a function $f_0(\omega, \alpha)$ and a value of the regularization parameter $\alpha = \alpha_{\text{op}}$ that minimizes the functional $\Phi(z_\alpha, z_T)$:

$$\Phi(z^0_{\alpha_{\text{op}}}, z_T) = \inf_{\begin{cases} f \in \mathcal{F}_f \\ \alpha > 0 \end{cases}} \Phi(z_\alpha, z_T).$$

Here,

$$z^0_{\alpha_{op}} = R_{f_0}(u, \alpha_{op}).$$

We shall say that the regularizing operator $R_{f_0}(u, \alpha_{op})$ corresponding to the function $f_0(\omega, \alpha)$ and to the value of the regularization parameter $\alpha = \alpha_{op}$ is **optimal on the family**, $\mathcal{F}_f$ and that α_{op} is the optimal value of the regularization parameter on that family. The regularized solution of equation (4.1.5) obtained with the aid of the optimal regularizing operator will be called the **optimal regularized solution.**

Sometimes, it is impossible to find the optimal stabilizing factor (algorithm) in the family $\mathcal{F}_f$ but possible to find a factor close to it (in some specified sense). For this reason, we are interested in

Problem 3. Estimate $\Phi(z_{\alpha_1,f_1}, z_{\alpha_2,f_2})$ under the condition $\rho_{\mathcal{F}}(f_1, f_2) \leqslant \delta$, where $z_{\alpha_1,f_1} = R_{f_1}(u, \alpha_1)$, $z_{\alpha_2,f_2} = R_{f_2}(u, \alpha_2)$, and $\rho_{\mathcal{F}}(f_1, f_2)$ is a metric for $\mathcal{F}_f$.

These problems and modifications of them will be examined in the present and later chapters.

§2. Deviation of the regularized solution from the exact one.

1. Let us return to equation (4.1.5). Suppose that $u(t) = u_T(t) + v(t)$, where $v(t)$ is an interference (noise). Let us suppose that $v(t)$ is a random function uncorrelated with the solution $z_T(t)$ that we are seeking and that its mathematical expectation is equal to $0: \overline{v(t)} = 0$.

The regularized solution can be written in the form

$$z_\alpha(t) = \frac{1}{2\pi}\int_{-\infty}^{\infty} \frac{f(\omega, \alpha)}{K(\omega)} u_T(\omega) \exp(-i\omega t)\, d\omega +$$

$$+ \frac{1}{2\pi}\int_{-\infty}^{\infty} \frac{f(\omega, \alpha)}{K(\omega)} v(\omega) \exp(-i\omega t)\, d\omega.$$

Consequently,

$$z_\alpha(t) - z_T(t) = \frac{1}{2\pi}\int_{-\infty}^{\infty}\{f(\omega, \alpha) - 1\}\frac{u_T(\omega)}{K(\omega)}\exp(-i\omega t)\,d\omega +$$

$$+\frac{1}{2\pi}\int_{-\infty}^{\infty}\frac{f(\omega, \alpha)}{K(\omega)}v(\omega)\exp(-i\omega t)\,d\omega. \quad (4.2.1)$$

The first term in the right-hand member of this equation characterizes the influence of the regularization (when the right-hand member is exact); the second characterizes the influence of the noise in the right-hand member of equation (4.1.5).

Let us define

$$\Delta_r(t, \alpha) = \frac{1}{2\pi}\int_{-\infty}^{\infty}\{f(\omega, \alpha) - 1\}\frac{u_T(\omega)}{K(\omega)}\exp(-i\omega t)\,d\omega =$$

$$= \frac{1}{2\pi}\int_{-\infty}^{\infty}\{f(\omega, \alpha) - 1\}z_T(\omega)\exp(-i\omega t)\,d\omega, \quad (4.2.2)$$

$$\Delta_n(t, \alpha) = \frac{1}{2\pi}\int_{-\infty}^{\infty}\frac{f(\omega, \alpha)}{K(\omega)}v(\omega)\exp(-i\omega t)\,d\omega, \quad (4.2.3)$$

so that (4.2.1) becomes

$$z_\alpha(t) - z_T(t) = \Delta_r(t, \alpha) + \Delta_n(t, \alpha).$$

It was shown in §1 (see theorem in subsection 5) that $\Delta_r(t, \alpha)$ approaches zero as $\alpha \to 0$ and that the convergence is uniform with respect to t. Since $v(t)$ is a random function, $\Delta_n(t, \alpha)$ is also a random function and its mathematical expectation is zero: $\overline{\Delta_n(t, \alpha)} = 0$.

Remark. If $f(t)$ is a random function (a random process), then the function

$$R(t_1, t_2) = \overline{f(t_1)\,f(t_2)}$$

is called the **autocorrelation function** of the process $f(t)$. For *stationary* random processes, the autocorrelation function depends only on the difference $t_2 - t_1$, so that we may write

$$R(t) = \overline{f(\tau)f(t+\tau)}.$$

The Fourier transform of the autocorrelation function of a stationary random process $f(t)$ is called its **spectral density** $S(\omega)$. It is well known* that, for real ω, the function $S(\omega)$ is a nonnegative even function such that $S(\omega) \leqslant S(0)$ and $S(\omega) \to 0$ as $\omega \to \infty$.

2. In what follows, we shall assume that $v(t)$ is the sample function of a stationary random process with spectral density $S(\omega)$. Under these assumptions, the variance of the random function $\Delta_n(t, \alpha)$ (the variance of the influence of the noise) is

$$\overline{\Delta_n^2(t, \alpha)} = \sigma^2(\alpha) = \frac{1}{4\pi^2} \int_{-\infty}^{\infty} \frac{f^2(\omega, \alpha)}{L(\omega)} S(\omega)\, d\omega, \qquad (4.2.4)$$

where

$$L(\omega) = K(\omega) K(-\omega).$$

It is obvious from property 6_f of the function $f(\omega, \alpha)$ that $\sigma^2(\alpha)$ increases as $\alpha \to \infty$. If

$$\frac{S(\omega)}{L(\omega)} \in L_1(-\infty, \infty),$$

then $\sigma^2(\alpha)$ approaches a finite value $\sigma^2(0)$ as $\alpha \to 0$. On the other hand, if

$$\frac{S(\omega)}{L(\omega)} \notin L_1(-\infty, \infty),$$

---

*A. A. Sveshnikov, *Prikladnyye metody teorii sluchaynykh funktsiy* (Applied methods in the theory of random functions), Moscow, Nauka press, 1968.

then $\sigma^2(\alpha) \to \infty$ as $\alpha \to 0$. It immediately follows from the properties of the function $f(\omega, \alpha)$ that $\sigma^2(\alpha) \to 0$ as $\alpha \to \infty$.

3. The metric in which we estimate the deviation of $z_\alpha(t)$ from $z_T(t)$ is chosen on the basis of the nature of the information available regarding the noise in the right-hand member of the equation. If $v(t)$ is a random function, it is natural to use a probabilistic metric.

Let us estimate the deviation between $z_\alpha(t)$ and $z_T(t)$ from the formula

$$\rho_F^2(z_\alpha, z_T) = \sup_t \overline{[z_\alpha(t) - z_T(t)]^2}.$$

Suppose that

$$\Delta_r^2(\alpha) = \sup_t \Delta_r^2(t, \alpha).$$

Then

$$T_f(\alpha) = \sup_t \overline{[z_\alpha(t) - z_T(t)]^2} = \Delta_r^2(\alpha) + \sigma^2(\alpha). \quad (4.2.5)$$

As $\alpha \to 0$, $\Delta_r^2(\alpha) \to 0$ (see p. 116) and $\sigma^2(\alpha)$ increases monotonically. Consequently, $T_f(\alpha)$ attains its smallest value at some value α_0 of α. (If this value is not unique, let us take the smallest such value.) We shall call α_0 the **(*C*, *f*)-optimal** value of the regularization parameter. The regularized solution obtained for $\alpha = \alpha_0$ is the asymptotically (as $\alpha \to 0$) best in the sense of $\min_\alpha T_f(\alpha)$ in the class of regularized solutions corresponding to a given function $f(\omega, \alpha)$.

It will be of interest to ascertain the classes of equations of the form (4.1.5) for which we can, by using the corresponding information regarding the solution sought and the noise, determine the (*C*, *f*)-optimal value of the regularization parameter α or a value close to it. The very definition of α_0 indicates that, to find it, we need to calculate $\Delta_r^2(\alpha)$ and $\sigma^2(\alpha)$ for small values of α; that is, we need to find asymptotic representations of the functions

$\Delta_r^2(t, \alpha)$, $\Delta_r^2(\alpha)$, and $\sigma^2(\alpha)$ as $\alpha \to 0$. Thus, the problem arises of asymptotic estimation, as $\alpha \to 0$, of $\Delta_r^2(t, \alpha)$, $\Delta_r^2(\alpha)$, and $\sigma^2(\alpha)$. Such estimates will be given in §3 for certain classes of equations.

4. At first glance, we might expect to have to find these estimates for each stabilizing factor $f(\omega, \alpha)$. However, it is possible to exhibit classes of stabilizing factors (see [14, 15]) such that the asymptotic estimates for $\Delta_r^2(t, \alpha)$, $\Delta_r^2(\alpha)$, and $\sigma^2(\alpha)$ as $\alpha \to 0$ will be the same for stabilizing factors in the same class. Therefore, it will be sufficient to obtain these estimates for the simplest stabilizing factors in the class in question.

Definition. We shall say that two stabilizing factors $f_1(\omega, \alpha_1)$ and $f_2(\omega, \alpha_2)$ are **asymptotically ϵ-close** if, for some $\epsilon > 0$, there exists an $\alpha_0(\epsilon)$ such that the inequality*

$$\left\| \frac{f_1(\omega, \alpha_1)}{K(\omega)} - \frac{f_2(\omega, \alpha_2)}{K(\omega)} \right\|_{L_2} \leqslant \epsilon$$

holds for all α_1 and α_2 less than $\alpha_0(\epsilon)$ such that $\left| \frac{\alpha_1}{\alpha_2} - 1 \right| \leqslant \min\{\alpha_1^2, \alpha_2^2\}$. If we take

$$f(\omega, \alpha) = \frac{L(\omega)}{L(\omega) + \alpha M(\omega)},$$

then the stabilizing factors

$$f_1(\omega, \alpha_1) = \frac{L(\omega)}{L(\omega) + \alpha_1 M_1(\omega)},$$

$$f_2(\omega, \alpha_2) = \frac{L(\omega)}{L(\omega) + \alpha_2 M_2(\omega)}$$

defined by the functions

*$\| \varphi_1(\omega) - \varphi_2(\omega) \|_{L_2}$ is the distance between the functions $\varphi_1(\omega)$ and $\varphi_2(\omega)$ in the metric for $L_2(-\infty, \infty)$.

$$M_1(\omega) = \omega^{2p}, \tag{4.2.6}$$

$$M_2(\omega) = \omega^{2p} + q_{p-1}\omega^{2p-2} + \ldots + q_0, \tag{4.2.7}$$

where

$$q_0, q_1, \ldots, q_{p-1} \geqslant 0,$$

are asymptotically ϵ-close for every $\epsilon > 0$.

5. Let us estimate the difference between the regularized solutions of equation (4.1.5) obtained with the aid of asymptotically ϵ-close stabilizing factors $f_1(\omega, \alpha_1)$ and $f_2(\omega, \alpha_2)$. Obviously,

$$\begin{aligned} z_{\alpha_1, f_1}(t) - z_{\alpha_2, f_2}(t) \,| &\leqslant \\ &\leqslant \frac{1}{2\pi} \int_{-\infty}^{\infty} \left| \frac{f_1(\omega, \alpha_1)}{K(\omega)} - \frac{f_2(\omega, \alpha_2)}{K(\omega)} \right| |u(\omega)| \, d\omega \leqslant \\ &\leqslant \frac{1}{2\pi} \left\| \frac{f_1(\omega, \alpha_1)}{K(\omega)} - \frac{f_2(\omega, \alpha_2)}{K(\omega)} \right\|_{L_2} \cdot \| u(t) \|_{L_2} \leqslant \epsilon \, \| u(t) \|_{L_2}. \end{aligned}$$

Since the stabilizing factors defined by the functions (4.2.6) and (4.2.7) are asymptotically ϵ-close for every $\epsilon > 0$, it will be sufficient, in the examination, for example, of regularized solutions obtained with the aid of pth-order stabilizers with constant coefficients, to obtain asymptotic estimates of the functions $\Delta_r^2(t, \alpha)$, $\Delta_r^2(\alpha)$, and $\sigma^2(\alpha)$ as $\alpha \to 0$ for regularized solutions obtained with the *simplest pth order stabilizers*, that is, for $M(\omega) = \omega^{2p}$.

§3. Asymptotic estimates of the deviation of a regularized solution from the exact solution for an equation of the convolution type as $\alpha \to 0$.

1. Following [9, 10], we shall obtain below asymptotic formulas for $\Delta_r(t, \alpha)$ and $\sigma^2(\alpha)$ for various types of equation of

the form (4.1.5). These types are determined by the asymptotic nature of the Fourier transform of the kernel $K(\omega)$ as $|\omega| \to \infty$. Here, we shall look at four types of equation.

Type I. $K(\omega)$ is a rational function that has no zeros on the real axis but that approaches zero as $\omega \to \infty$ at the same rate as

$$\frac{A}{\omega^n},$$

where n is a positive integer.

Type II. All finite singular points of the function $K(\omega)$ are located in a bounded region. For every positive number B, the integral

$$\int_0^B \frac{d\omega}{L(\omega)}$$

converges. As $\omega \to \infty$,

$$K(\omega) \approx H \exp[-(iA\omega)^{\frac{1}{m}}],$$

where $m > 1$ and $A > 0$.

Type III. As $\omega \to \infty$,

$$K(\omega) \approx H \exp(-A\omega^2),$$

where H and A are positive.

Type IV. As $|\omega| \to \infty$,

$$K(\omega) \approx \frac{A}{\omega^n (\underbrace{\ln \ln \ldots \ln}_{s \text{ times}} |\omega|)^k},$$

where A is a positive number, and n and s are nonnegative integers. If $n > 0$, there is no restriction on k except that it be an integer, but if $n = 0$, both k and s are positive.

2. Various problems reduce to equations with kernels of these types. As examples, the problem of calculating the nth derivative, the problem of automatic control, and the problem of restoring electromagnetic signals distorted by a filter with concentrated parameters (capacitance, inductance, resistance) lead to kernels of the first type; problems of restoring electromagnetic signals distorted by a long line with losses (a cable or a conducting medium) lead to the second type (with $m = 2$); problems of restoring electromagnetic signals propagated over a spherical surface received in a region of geometric shadow also lead to the second type; problems of restoring acoustic signals propagated through a viscous medium lead to the third type. There are many others.

3. We shall assume that the spectral density of the noise $S(\omega)$, for real ω, is such that, as $|\omega| \to \infty$,

$$S(\omega) = \frac{S_0}{\omega^a},$$

where a is a nonnegative and S_0 a positive constant. If $a = 0$ and $S(\omega) = S_0$, we have a "white" noise. Here, S_0 is a characteristic of the noise level.

In what follows, we shall examine the asymptotic behavior of the functions $\Delta_r(t, \alpha)$ and $\sigma^2(\alpha)$ only for stabilizers with constant coefficients: $q_i(s) = q_i = \text{const}$. In this case,

$$M(\omega) = \omega^{2p} + q_{p-1}\omega^{2p-2} + \ldots + q_0$$

(for a pth-order stabilizer).

Since the stabilizing factors $L(\omega)/[L(\omega) + \alpha M(\omega)]$ defined by the functions

$$M_1(\omega) = \omega^{2p} + q_{p-1}\omega^{2p-2} + \ldots + q_0,$$

$$M_2(\omega) = \omega^{2p},$$

are asymptotically ϵ-close for every $\epsilon > 0$, it will, by virtue of §2 of the present chapter, be sufficient to get asymptotic estimates of the functions $\Delta_r(t, \alpha)$ and $\sigma^2(\alpha)$ for the functions $M(\omega) = \omega^{2p}$.

4. Estimation of $\Delta_r(t, \alpha)$ for equations with kernel of the first type. For a pth-order stabilizer for which $M(\omega) = \omega^{2p}$, the regularized solution of equation (4.1.5) with exact right-hand member $u_T(t)$ has the form

$$z_\alpha(t) = \frac{1}{2\pi}\int_{-\infty}^{\infty} \frac{L(\omega)\, z_T(\omega)}{L(\omega) + \alpha\omega^{2p}} \exp(-i\omega t)\, d\omega$$

or

$$z_\alpha(t) = \int_{-\infty}^{\infty} L_\alpha(t-\tau)\, z_T(\tau)\, d\tau = \int_{-\infty}^{\infty} L_\alpha(\xi) \cdot z_T(t-\xi)\, d\xi, \tag{4.3.1}$$

where

$$L_\alpha(t) = \frac{1}{2\pi}\int_{-\infty}^{\infty} \frac{L(\omega)\exp(-i\omega t)}{L(\omega) + \alpha\omega^{2p}}\, d\omega \tag{4.3.2}$$

is an even function. Therefore,

$$z_\alpha(t) = \int_{0}^{\infty} L_\alpha(\xi)\{z_T(t-\xi) + z_T(t+\xi)\}\, d\xi. \tag{4.3.3}$$

The integral in (4.3.2) is equal to the sum of the residues of the integrand at its poles (multiplied by $2\pi i$) located in the upper half-plane for $t < 0$ and in the lower half-plane for $t > 0$. The poles of the integrand are the zeros of the denominator $L(\omega) + \alpha\omega^{2p}$, which are of one or the other of two kinds:

1) zeros $\omega_{1\alpha}$ that approach the zeros of the function $L(\omega)$ as $\alpha \to 0$,

2) zeros $\omega_{2\alpha}$ that become infinite as $\alpha \to 0$. Let us look at the contribution of each of these to $L_\alpha(t)$.

For zeros of the first kind, $\omega_{1\alpha} \to \omega_1$ as $\alpha \to 0$. If ω_1 is a simple zero of the function $L(\omega)$, we have

$$\operatorname{res}\left\{\frac{L(\omega)\exp(-i\omega t)}{L(\omega)+\alpha\omega^{2p}}\right\}_{\omega=\omega_{1\alpha}} = \frac{L(\omega_{1\alpha})\exp(-i\omega_{1\alpha}t)}{2p\alpha\omega_{1\alpha}^{2p-1}+\left(\dfrac{dL}{d\omega}\right)_{\omega=\omega_{1\alpha}}} =$$

$$= \frac{(\omega_{1\alpha}-\omega_1)\cdot\left(\dfrac{dL}{d\omega}\right)_{\omega=\omega_1}}{2p\alpha\omega_1^{2p-1}+\left(\dfrac{dL}{d\omega}\right)_{\omega=\omega_1}}\exp(-i\omega_1 t) + O[(\omega_{1\alpha}-\omega_1)^2] =$$

$$= (\omega_{1\alpha}-\omega_1)\exp(-i\omega_1 t)[1+O(\alpha)] + O[(\omega_{1\alpha}-\omega_1)^2].$$

Since

$$f(\omega) = L(\omega) + \alpha\omega^{2p} =$$

$$= f(\omega_1) + (\omega-\omega_1)\left(\frac{df}{d\omega}\right)_{\omega=\omega_1} + O[(\omega_1-\omega)^2],$$

$$f(\omega_{1\alpha}) = 0 \text{ and } L(\omega_1) = 0,$$

it follows that

$$0 = \alpha\omega_1^{2p} + \left[\left(\frac{dL}{d\omega}\right)_{\omega=\omega_1} + 2p\alpha\omega_1^{2p-1}\right](\omega_{\alpha}-\omega_1) +$$

$$+ O[(\omega_{1\alpha}-\omega_1)^2].$$

Therefore,

$$\omega_{1\alpha}-\omega_1 = \frac{-\alpha\omega_1^{2p}}{\left(\dfrac{dL}{d\omega}\right)_{\omega=\omega_1}}\{1+O(\alpha)+O[(\omega_{1\alpha}-\omega_1)^2]\}.$$

Consequently,

$$\omega_{1\alpha} - \omega_1 = \frac{-\alpha\omega_1^{2p}}{\left(\frac{dL}{d\omega}\right)_{\omega=\omega_1}} \{1 + O(\alpha)\},$$

so that

$$\operatorname{res}\left\{\frac{L(\omega)\exp(-i\omega t)}{L(\omega)+\alpha\omega^{2p}}\right\}_{\omega=\omega_{1\alpha}} = \frac{-\alpha\omega_1^{2p}}{\left(\frac{dL}{d\omega}\right)_{\omega=\omega_1}} \times$$

$$\times \exp(-i\omega_1 t)\,[1 + O(\alpha)] = O(\alpha)\exp(-i\omega_1 t).$$

If ω_1 is a zero of multiplicity γ, we find by analogous reasoning (under the assumption that $L(\omega)$ is $\gamma - 1$ times differentiable) that

$$\operatorname{res}\left\{\frac{L(\omega)\exp(-i\omega t)}{L(\omega)+\alpha\omega^{2p}}\right\}_{\omega=\omega_{1\alpha}} = O(\alpha^{1/\gamma})\exp(-i\omega_1 t).$$

The number of zeros of the first kind is finite and independent of α. Therefore, the contribution of poles of the first kind $\omega_{1\alpha}$ to the function $L_\alpha(t)$ is $O(\alpha^{1/\gamma_0})$, where γ_0 is the highest multiplicity of any of the zeros of the function $L(\omega)$.

5. Let us now look at the contribution of poles of the second kind $\omega_{2\alpha}$. Since $\omega_{2\alpha} \to \infty$ as $\alpha \to 0$, we can, for sufficiently small α, use the asymptotic representation of the function $L(\omega)$; that is, we can set $L(\omega) = |A|^2/\omega^{2n}$. Therefore,

$$\operatorname{res}\left\{\frac{L(\omega)\exp(-i\omega t)}{L(\omega)+\alpha\omega^{2p}}\right\}_{\omega=\omega_{2\alpha}} = \frac{|A|^2\exp(-i\omega_{2\alpha}t)\cdot\omega_{2\alpha}}{2q\alpha\omega_{2\alpha}^{2q}},$$

$$q = n + p.$$

Thus, the contribution $L_{2\alpha}(t)$ of poles of the second kind to $L_\alpha(t)$ is, for $t \leqslant 0$,

$$L_{2\alpha}(t) = -\sum \frac{|A|^2 i\omega_{2\alpha} \exp(-i\omega_{2\alpha} t)}{2q\alpha\omega_{2\alpha}^{2q}},$$

where the summation is over all roots of the second kind of the equation $L(\omega) + \alpha\omega^{2p} = 0$ that lie in the upper half-plane. For small α, these roots can be replaced with the roots $\omega_\alpha^{(k)}$ of the equation

$$1 + \frac{\alpha}{|A|^2}\omega^{2q} = 0. \tag{4.3.4}$$

Therefore,

$$L_{2\alpha}(t) = -\sum_k \frac{i\omega_\alpha^{(k)} \exp(-i\omega_\alpha^{(k)} t)}{2q} \tag{4.3.5}$$

for $t \leqslant 0$. Analogously,

$$L_{2\alpha}(t) = \sum_k \frac{i\omega_\alpha^{(k)} \exp(-i\omega_\alpha^{(k)} t)}{2q} \tag{4.3.6}$$

for $t \geqslant 0$. Here, the summations are over all roots of equation (4.3.4) that are located in the upper (resp. lower) half-plane for $t \leqslant 0$ (resp. $t \geqslant 0$). As $\alpha \to 0$, the roots $\omega_\alpha^{(k)}$ approach infinity along the rays

$$\arg \omega = \frac{2k+1}{2q}\pi \quad (k \leqslant n).$$

Also,

$$|\omega_\alpha^{(k)}| = \left(\frac{|A|^2}{\alpha}\right)^{\frac{1}{2q}}.$$

We note that $n + p$ roots of equation (4.3.4) lie in the upper half-plane and $n + p$ in the lower.

If in formula (4.3.3) we replace $L_\alpha(t)$ with $L_{2\alpha}(t)$, we obtain from formula (4.3.6), for $t \geqslant 0$,

$$z_\alpha(t) = \int_0^\infty \sum_{k=0}^{q} \frac{+i\omega_\alpha^{(k)} \exp(-i\omega_\alpha^{(k)}\xi)}{2q} \{z_T(t+\xi) + z_T(t-\xi)\}\, d\xi \tag{4.3.7}$$

or

$$z_\alpha(t) = \frac{1}{2q} \sum_{k=0}^{q} \int_0^\infty \{-i\omega_\alpha^{(k)} \exp(-i\omega_\alpha^{(k)}\xi)[z_T(t+\xi) + z_T(t-\xi)]\} d\xi. \tag{4.3.8}$$

To estimate these integrals, we use the asymptotic formula

$$\int_0^\infty f(t \pm \xi)\, \beta \exp(-\beta\xi)\, d\xi = f(t) + \frac{1}{\beta} f'(t) + O\left(\frac{1}{\beta^2}\right),$$

which is valid for large values of $\operatorname{Re} \beta > 0$, as can be established by twice integrating by parts. Let us set $\beta = i\omega_\alpha^{(k)}$. Since the summation in formula (4.3.7) is over only those roots of equation (4.3.4) in the lower half-plane, we have $\operatorname{Re}(i\omega_\alpha^{(k)}) > 0$. If we set $f(t) = z_T(t)$, we obtain

$$z_\alpha(t) = \frac{1}{2q} \sum_{k=0}^{q} \{2z_T(t) + \frac{2i}{\omega_\alpha^{(k)}} z_T'(t)\} + O([\omega_\alpha^{(k)}]^{-2}).$$

Since

$$\omega_\alpha^{(k)} = \left(\frac{|A|^2}{\alpha}\right)^{\frac{1}{2q}} e^{i\frac{2k+1}{2q}\pi},$$

we have*

$$z_\alpha(t) = z_T(t) + \frac{1}{q}\left(\frac{|A|^2}{\alpha}\right)^{\frac{-1}{2q}} \frac{z_T'(t)}{\sin\left(\frac{\pi}{2q}\right)}[1 + O(\alpha^{\frac{1}{q}})].$$

Here, we used the relationships

$$\sum_{k=0}^{q} \cos\left(\frac{2k+1}{2q}\pi\right) = -\cos\frac{\pi}{2q}, \qquad \sum_{k=0}^{q} \sin\left(\frac{2k+1}{2q}\pi\right) = \frac{\cos^2\left(\frac{\pi}{2q}\right)}{\sin\left(\frac{\pi}{2q}\right)}.$$

Thus, we have the asymptotic (as $\alpha \to 0$) formula

$$\Delta_r^2(t, \alpha) = \frac{1}{q^2}\sin^{-2}\left(\frac{\pi}{2q}\right)\left(\frac{\alpha}{|A|^2}\right)^{\frac{1}{q}}\left|\frac{dz_T}{dt}\right|^2[1 + O(\alpha^{\frac{1}{q}})]. \tag{4.3.9}$$

Therefore, we have

Theorem 1. The asymptotic (as $\alpha \to 0$) formula (4.3.9) is valid in the use of pth-order regularization for equation (4.1.5) with a kernel of the first kind [10].

In particular, this formula is valid for stabilizers of zeroth order, that is, for $p = 0$.

We know [185] that, in the use of zeroth-order stabilizers, the regularized solution $z_\alpha(t)$ of an equation with right-hand member $u_T(t)$ will not in general approximate the exact solution $z_T(t)$ uniformly on $(-\infty, \infty)$ as $\alpha \to 0$. From formula (4.3.9) with $p = 0$, we obtain the

Corollary. *If the derivative of the solution $z_T(t)$ of equation* (4.1.5) *with a kernel of the first kind and with right-hand member*

---

*We assume that $\gamma_0 < 2q$.

$u = u_T(t)$ is bounded on $(-\infty, \infty)$, then the regularized solution $z_\alpha(t)$ of equation (4.1.5) obtained with the aid of a pth-order stabilizer (where $p = 0, 1, 2, \ldots$) approximates uniformly on $(-\infty, \infty)$ the exact solution $z_T(t)$ as $\alpha \to 0$.

6. An estimate of $\Delta_r(t, \alpha)$ for equations with kernels of types II–IV.

Theorem 2. In the use of pth-order stabilizers with constant coefficients for obtaining a regularized solution of equation (4.1.5) with type-II kernel, the following asymptotic (as $\alpha \to 0$) formulas are valid [10]:

$$\Delta_r(t, \alpha) = \sqrt{\frac{2}{\pi}}\, A\left(\frac{C_m}{\beta_2}\right)^m z_T'(t)\left[1 + O\left(\frac{1}{\beta_2}\right)\right], \tag{4.3.10}$$

$$\Delta_r(t, \alpha) = \frac{A^{2p+1}}{2p+1}\sqrt{\frac{2}{\pi}}\left(\frac{C_m}{\beta_2}\right)^{(2p+1)m}\frac{d^{2p+1}z_T}{dt^{2p+1}}\left[1 + O\left(\frac{1}{\beta_2}\right)\right], \tag{4.3.11}$$

where

$$C_m = 2\cos\left(\frac{\pi}{2m}\right),\quad \beta_2 = \ln\left(\frac{H^2A^{2p}}{\alpha}\right).$$

Theorem 3. *In the use of pth-order stabilizers with constant coefficients for obtaining a regularized solution of equation* (4.1.5) *with type-III kernel, the following asymptotic (as $\alpha \to 0$) formulas are valid* [10]:

$$\Delta_r(t, \alpha) = \sqrt{\frac{2}{\pi}}\left(\frac{2A}{\beta_3}\right)^{1/2}\frac{dz_T}{dt}\left[1 + O\left(\frac{1}{\beta_3}\right)\right], \tag{4.3.12}$$

$$\Delta_r(t, \alpha) = \sqrt{\frac{2}{\pi}}\,\frac{1}{2p+1}\left(\frac{2A}{\beta_3}\right)^{\frac{2p+1}{2}}\frac{d^{2p+1}z_T}{dt^{2p+1}}\left[1 + O\left(\frac{1}{\beta_3}\right)\right], \tag{4.3.13}$$

where

$$\beta_3 = \ln\left[\frac{2^p A^p H^2}{\alpha}\right].$$

Theorem 4. *In the use of pth-order stabilizers with constant coefficients for obtaining a regularized solution of equation* (4.1.5) *with type IV kernel, the following asymptotic (as $\alpha \to 0$) formulas are valid* [10]:

$$\Delta_r(t, \alpha) = \frac{1}{q \sin\left(\frac{\pi}{2q}\right)} \left(\frac{\alpha}{A^2} \beta_4^{2k}\right)^{\frac{1}{2q}} \frac{dz_T}{dt}\left[1 + O\left(\frac{1}{\beta_4}\right)\right], \tag{4.3.14}$$

$$\Delta_r(t, \alpha) = \frac{1}{q \sin\left(\frac{2p+1}{2q}\pi\right)} \left(\frac{\alpha}{A^2} \beta_4^{2k}\right)^{\frac{2p+1}{2q}} \frac{d^{2p+1} z_T}{dt^{2p+1}}\left[1 + O\left(\frac{1}{\beta_4}\right)\right], \tag{4.3.15}$$

where

$$\beta_4 = \underbrace{\ln\ln\ldots\ln}_{s \text{ times}}\left[\left(\frac{A^2}{\alpha}\right)^{\frac{1}{2q}}\right].$$

The formulas for $\Delta_r(t, \alpha)$ in Theorems II–IV are proven, in a manner analogous to the proof of Theorem 1, by evaluating the corresponding integrals with the aid of residues though the calculations are more laborious (see [10, 13]), and for this reason we shall not give the proofs here.

7. An estimate of $\sigma^2(\alpha)$ for equations with kernels of types I–IV. For

$$f(\omega, \alpha) = \frac{L(\omega)}{L(\omega) + \alpha M(\omega)},$$

formula (4.2.4) takes the form

$$\sigma^2(\alpha) = \frac{1}{2\pi^2} \int_0^\infty \frac{L(\omega) S(\omega) d\omega}{\{L(\omega) + \alpha M(\omega)\}^2}. \qquad (4.3.16)$$

It follows immediately from this formula that $\sigma^2(\alpha) \to \infty$ as $\alpha \to 0$ for equations with kernels of types I–IV (for types I and IV, we set $n > \max(\alpha, \frac{1}{2})$). On the other hand, for any fixed $\alpha > 0$, the integral (4.3.16) converges. Consequently, for sufficiently small values of the parameter α, the basic contribution to the integral (4.3.16) is provided by large frequencies ω. Keeping this in mind, let us represent the integral (4.3.16) as the sum of two integrals:

$$\sigma^2(\alpha) = \frac{1}{2\pi^2} \int_0^{\omega_0} \ldots + \frac{1}{2\pi^2} \int_{\omega_0}^\infty \ldots .$$

Let us take ω_0 sufficiently large that, for $\omega \geqslant \omega_0$, the functions $L(\omega)$ and $S(\omega)$ can be replaced with their asymptotic representations. Then, for $M(\omega) = \omega^{2p}$, we have

$$\sigma^2(\alpha) = \frac{1}{2\pi^2} \int_0^{\omega_0} \left\{ \frac{L \cdot S}{(L + \alpha\omega^{2p})^2} - \frac{L_{as} \cdot S_{as}}{(L_{as} + \alpha\omega^{2p})^2} \right\} d\omega +$$
$$+ \frac{1}{2\pi^2} \int_0^\infty \frac{L_{as} \cdot S_{as}\, d\omega}{(L_{as} + \alpha\omega^{2p})^2}.$$

Taking $S(0) = D \cdot S_0$, where $D > 0$, and remembering that $S(\omega) \leqslant S(0)$ for every ω, we immediately find that, as $\alpha \to 0$,

$$\frac{1}{2\pi^2} \int_0^{\omega_0} \left\{ \frac{L \cdot S}{(L + \alpha\omega^{2p})^2} - \frac{L_{as} S_{as}}{(L_{as} + \alpha\omega^{2p})^2} \right\} d\omega = S_0 \cdot O(1).$$

Here, $S_{as}(\omega)$ and $L_{as}(\omega)$ are the values of the functions $S(\omega)$ and $L(\omega)$ calculated from their asymptotic formulas. Consequently,

$$\sigma^2(\alpha) = S_0 \cdot O(1) + I_1 \tag{4.3.17}$$

and the problem reduces to estimating the limit (as $\alpha \to 0$) of the integral

$$I_1 = \frac{1}{2\pi^2} \int_0^{\infty} \frac{L_{as} \cdot S_{as}\, d\omega}{(L_{as} + \alpha\omega^{2p})^2}. \tag{4.3.18}$$

For equations with type-I kernels, the integral I_1 has the form

$$I_1 = \frac{1}{2\pi^2} \int_0^{\infty} \frac{S_0 \omega^{2n-2a}\, d\omega}{|A|^2\left(1 + \frac{\alpha}{|A|^2}\omega^{2q}\right)^2}$$

and, when we make the change of variable $x = (\alpha/|A^2|)\omega^{2q}$, it can be evaluated exactly. We obtain

$$I_1 = \frac{1}{4q\pi^2} \frac{S_0}{|A|^2} \left(\frac{|A|^2}{\alpha}\right)^{\frac{2n-2a+1}{2q}} \frac{2p + 2a - 1}{2q \sin\left(\frac{2p + 2a - 1}{2q}\pi\right)}. \tag{4.3.19}$$

Thus, we have

Theorem 5. *For equations with type-I kernel, the variance of the influence of the noise in a regularized solution* $\sigma^2(\alpha)$ *is calculated from formulas* (4.3.17) *and* (4.3.19) *and it approaches infinity as* $\alpha \to 0$.

8. For equations with type-II kernels, we have

$$\sigma^2(\alpha) = \frac{1}{2\pi^2} \int_0^{\omega_0} \frac{L \cdot S\, d\omega}{(L + \alpha\omega^{2p})^2} + \frac{1}{2\pi^2} \int_{\omega_0}^{\infty} \frac{L_{as} \cdot S_{as}\, d\omega}{(L_{as} + \alpha\omega^{2p})^2},$$

where ω_0 is chosen as in subsection 1.

Obviously, as $\alpha \to 0$,

$$\int_0^{\omega_0} \frac{L \cdot S\, d\omega}{(L + \alpha\omega^{2p})^2} = S_0 \cdot O(1).$$

Thus,

$$\sigma^2(\alpha) = S_0 \cdot O(1) + I_2, \tag{4.3.20}$$

where

$$I_2 = \frac{1}{2\pi^2} \int_{\omega_0}^{\infty} \frac{L_{as}\; S_{as}\, d\omega}{(L_{as} + \alpha\omega^{2p})^2}\,.$$

We can write the integrand in the expression for I_2 in the form $\omega^{-2a} F(\omega, \alpha)$. The function $F(\omega, \alpha)$ has a sharply expressed maximum whose position ω_1 approaches infinity as $\alpha \to 0$. We can use the method of steepest descent to estimate I_2. We find ω_1 from the equation $\partial F(\omega, \alpha)/\partial\omega = 0$, which can be written in the form

$$\alpha\omega^{2p} K_1'(\omega) - K_1'(\omega) K_1^2(\omega) - 2p\omega^{2p-1} K_1(\omega) = 0, \tag{4.3.21}$$

where

$$K_1(\omega) = H \exp\left[-0.5\, C_m (A\omega)^{\frac{1}{m}}\right] = \{L(\omega)\}^{1/2}.$$

Since, for large values of ω,

$$\omega K_1'(\omega) \gg K_1(\omega),$$

we can replace equation (4.3.21) with the equation

$$L(\omega) = \alpha\omega^{2p}. \tag{4.3.22}$$

One can easily see that the formula

$$\omega_1 = \frac{1}{A}\left(\frac{\beta_2}{C_m}\right)^m\left[1 + O\left(\frac{1}{\beta_2}\right)\right] \tag{4.3.23}$$

holds when equation (4.3.22) has a unique positive root ω_1.

Applying the method of steepest descent to evaluate the integral I_2, we obtain

$$I_2 = \frac{mS_0 \cdot}{4\sqrt{\pi^3}C_m} \frac{\omega_1^{1-2p-2a}}{\alpha(A\omega_1)^{\frac{1}{m}}},$$

which, together with formula (4.3.23), yields

$$I_2 = \frac{mS_0}{4C_m\sqrt{\pi^3}} \frac{A^{2p+2a-1}}{\alpha}\left(\frac{C_m}{\beta_2}\right)^{2p+2a-1+\frac{1}{m}}\left[1 + O\left(\frac{1}{\beta_2}\right)\right]. \tag{4.3.24}$$

Thus, we have

Theorem 6. *For equations of the form* (4.1.5) *with type-II kernel, the variance of the influence of the noise* $\sigma^2(\alpha)$ *in a regularized solution obtained with the aid of the simplest pth-order stabilizer* ($M(\omega) = \omega^{2p}$) *is calculated from formulas* (4.3.20) *and* (4.3.24) *and it approaches infinity at the same rate as* I_2 *as* $\alpha \to 0$.

9. In an analogous manner, we can establish the following theorems:

Theorem 7. *For equations of the form (4.1.5) with type-III kernel, the variance of the influence of the noise* $\sigma^2(\alpha)$ *in a regularized solution obtained by means of the simplest pth-order stabilizer is calculated according to the formula*

$$\sigma^2(\alpha) = S_0 \cdot O(1) + I_3, \tag{4.3.25}$$

where

$$I_3 = \frac{S_0 (2A)^{2p+2a}}{8\sqrt{\pi^3} \cdot \alpha} \beta_3^{-2p-2a-1} \left[1 + O\left(\frac{1}{\beta_3}\right)\right]. \quad (4.3.26)$$

Theorem 8. *For equations of the form* (4.1.5) *with type-IV kernel, the variance of the influence of the noise* $\sigma^2(\alpha)$ *in a regularized solution obtained with the aid of the simplest pth-order stabilizer is calculated according to the formula*

$$\sigma^2(\alpha) = S_0 \cdot O(1) + I_4, \quad (4.3.27)$$

where

$$I_4 = \frac{\left(1 - \frac{1}{2q}\right) S_0 A^{\frac{a}{q}}}{2q \sin\left(\frac{2p-1}{2q}\pi\right)} \left(\frac{A}{\alpha}\right)^{\frac{2n-2a+1}{2q}} \beta_4^{k\frac{2p+2a-1}{2q}} \left[1 + O\left(\frac{1}{\beta_4}\right)\right].$$

(4.3.28)

10. Using formulas (4.3.9)–(4.3.15), (4.3.17), (4.3.19), (4.3.20), (4.3.24)–(4.3.28), we can find from the condition

$$\Delta_r^2(\alpha) = \sigma^2(\alpha)$$

a value of α that is close to the (C, f)-optimal value. We shall call this value **almost optimal** and shall denote it by α_{ao}. To find α_{ao}, we need to know the least upper bound of $|z_T'(t)|$ or $|d^{2p+1} z_T / dt^{2p+1}|$.

Suppose, for example that $\sup_t |z_T'(t)| = B_0$. Then, from formulas (4.3.9) and (4.3.19), we find

$$\alpha_{ao} = |A|^2 \left\{\frac{S_0}{4\pi^2 B_0^2} \frac{2p+2a-1}{|A|^2} \frac{\sin^2\left(\frac{\pi}{2q}\right)}{\sin\left(\frac{2p+2a-1}{2q}\pi\right)}\right\}^{\frac{2q}{2n-2a+3}}.$$

Remark 1. If we have in advance some information regarding the least upper bound of the absolute value of the first derivative of the exact solution $|z'_T(t)|$, then, to determine α_{ao}, we should use formulas (4.3.9), (4.3.10), (4.3.12), and (4.3.14) and accordingly formulas (4.3.19), (4.3.24), (4.3.26), and (4.3.28). If we have some information regarding the absolute value of the $(2p_0 + 1)$st derivative, to solve the problem we need to use a p_0th-order stabilizer, and to determine α_{ao}, we need to use relationships obtained from the formula

$$\sigma^2(\alpha) = \Delta_r^2(\alpha)$$

by using the relationships (4.3.9), (4.3.11), (4.3.13), and (4.3.15) and accordingly (4.3.19), (4.3.24), (4.3.26), and (4.3.28).

Remark 2. For some of the cases examined in this chapter, other estimates of the deviation of the regularized solution from the exact one when $M(\omega) = \omega^2$ are given in [4].

An examination is made in [79] of the asymptotic estimates of the deviation of the regularized solution from the exact one for a one-dimensional integral equation of the form (I.1.1) that is not of the convolution type.

CHAPTER V

CERTAIN OPTIMAL REGULARIZING OPERATORS FOR INTEGRAL EQUATIONS OF THE CONVOLUTION TYPE

1. Suppose that the right-hand member $u(t)$ of the equation

$$Az \equiv \int_{-\infty}^{\infty} K(t-\tau)\, z(\tau)\, d\tau = u(t) \tag{5.0.1}$$

contains a random interference (noise) $v(t)$ and hence is a random function. The regularized solutions of equation (5.0.1)

$$z_\alpha(t) = R_f(u, \alpha)$$

are also random functions. Just as in Chapter IV, we shall estimate their deviations from the exact solution $z_T(t)$ (or each other) in a probabilistic metric:

$$\rho_F(z_\alpha, z_T) = \overline{(z_\alpha - z_T)^2}, \tag{5.0.2}$$

where the vinculum denotes the mathematical expectation.

Having thus defined a metric in F, let us look at the regularized solutions obtained with the aid of stabilizing factors of the form

$$f(\omega, \alpha) = \frac{L(\omega)}{L(\omega) + \alpha M(\omega)}.$$

2. In the present chapter, we shall find, out of all such regularized solutions, the one that is optimal in the sense of the metric (5.0.2). It will be shown that the operator representing optimal Wiener filtering is a regularizing operator of the class described in Chapter IV. For equations with kernels of types I and II (see Chapter IV, §3), we shall find the optimal regularized solution in the class of solutions obtained with the aid of the simplest pth-order stabilizers ($M(\omega) = \omega^{2p}$) and we shall study the deviations of nonoptimal solutions from the optimal one [11, 14].

3. It should be mentioned that in the examination of equations of the form $Az = u$, where z is the unknown function depending on the variable t, two formulations of the problem are possible.

With the first formulation, the function $u = u_T(t)$ is deterministic and we are required to find a deterministic solution $z_T(t)$. If instead of the function $u_T(t)$; we know only a δ-approximation $u(t) = u_T(t) + v(t)$ of it such that $\rho_U(u, u_T) \leqslant \delta$, we can speak only of finding a solution that approximates $z_T(t)$. The interference $v(t)$ is usually a random variable.

With the second formulation, $u_T(t)$ is the sample function of a random process and we need to find the sample function of another random process $z_T(t)$ connected with the first by $Az_T = u_T$. If instead of $u_T(t)$, we have $u(t) = u_T(t) + v(t)$, where $v(t)$ is an interference (noise) in the form of a random process, then we seek a "solution" close to $z_T(t)$.

In solving this problem, the use of supplementary information regarding the solution sought and the noise is significant. Cases are possible in which

a) we know the spectral densities of the solution and of the noise,

b) we know the probability distributions of the solution and of the noise.

In the present chapter, the above-mentioned optimal regularized solution and the estimates of the deviation of a nonoptimal regularized solution from the optimal one will be treated as applied to the second formulation of the problem in the use of *a priori* information of type a). This will be done under the assumption that $v(t)$ and the solution sought are sample functions of two uncorrelated random processes. We shall denote by $S(\omega)$ and $N(\omega)$ respectively the spectral densities of these processes.

We shall indicate a method of finding approximate asymptotic representations of $S(\omega)$ and $N(\omega)$ as $\omega \to \infty$.

Considerations associated with the use of *a priori* information of type b) may be found in [179–181].

§1. The optimal regularized solution. The connection between the regularization method and optimal Wiener filtering.

1. Suppose that $z_T(t)$ is the exact solution of equation (5.0.1) with right-hand member $u = u_T(t)$; that is, $Az_T \equiv u_T(t)$. Let us suppose that $u(t) = u_T(t) + v(t)$. Let us look at regularized solutions of equation (5.0.1) of the form

$$z_\alpha(t) = \frac{1}{2\pi} \int_{-\infty}^{\infty} \frac{K(-\omega)\, u(\omega)}{L(\omega) + \alpha M(\omega)} \exp(-i\omega t)\, d\omega = R_M(u, \alpha), \tag{5.1.1}$$

where the function $\dfrac{K(-\omega)}{L(\omega) + \alpha M(\omega)}$ belongs to $L_2(-\infty, \infty)$ for every $\alpha > 0$. The operator $R_M(u, \alpha)$ is defined in terms of the function $M(\omega)$. It can be written in the form of a convolution

$$R_M(u, \alpha) = \int_{-\infty}^{\infty} L_\alpha(t - \tau)\, u(\tau)\, d\tau,$$

where

$$L_\alpha(t) = \frac{1}{2\pi}\int_{-\infty}^{\infty} \frac{K(-\omega)}{L(\omega)+\alpha M(\omega)} \exp(-i\omega t)\, d\omega.$$

2. Among operators of this kind, one can find an operator $R_{M_0}(u, \alpha)$ minimizing $\overline{[z_\alpha(t) - z_T(t)]^2}$. Obviously,

$$z_\alpha(t) - z_T(t) = \frac{1}{2\pi}\int_{-\infty}^{\infty} \left\{\frac{K(-\omega)\,u(\omega)}{L(\omega)+\alpha M(\omega)} - z_T(\omega)\right\} \exp(-i\omega t)\, d\omega =$$

$$= \frac{1}{2\pi}\int_{-\infty}^{\infty} \left\{\frac{K(-\omega)\,[u_T(\omega)+v(\omega)]}{L(\omega)+\alpha M(\omega)} - z_T(\omega)\right\} \exp(-i\omega t)\, d\omega =$$

$$= \frac{1}{2\pi}\int_{-\infty}^{\infty} \left\{\frac{L(\omega)\,z_T(\omega)+K(-\omega)\,v(\omega)}{L(\omega)+\alpha M(\omega)} - z_T(\omega)\right\} \exp(-i\omega t)\, d\omega =$$

$$= \frac{1}{2\pi}\int_{-\infty}^{\infty} \left\{\frac{-\alpha M(\omega)\,z_T(\omega)}{L(\omega)+\alpha M(\omega)} + \frac{K(-\omega)\,v(\omega)}{L(\omega)+\omega M(\omega)}\right\} \exp(-i\omega t)\, d\omega$$

since $u_T(\omega) = K(\omega) z_T(\omega)$. Therefore,

$$\overline{[z_\alpha(t) - z_T(t)]^2} =$$

$$= \overline{\frac{1}{2\pi}\int_{-\infty}^{\infty} \frac{-\alpha M(\omega)\,z_T(\omega)+K(-\omega)\,v(\omega)}{L(\omega)+\alpha M(\omega)} \exp(-i\omega t)\, d\omega} \times$$

$$\times \overline{\frac{1}{2\pi}\int_{-\infty}^{\infty} \frac{-\alpha M(\omega')\,z_T(\omega')+K(-\omega')\,v(\omega')}{L(\omega')+\alpha M(\omega')} \exp(-i\omega' t)\, d\omega'} =$$

$$= \frac{1}{4\pi^2}\int_{-\infty}^{\infty}\!\!\int \frac{\alpha^2 M(\omega)\,M(\omega')\,\overline{z_T(\omega)\,z_T(\omega')} + K(-\omega)\,K(-\omega')\,\overline{v(\omega)v(\omega')}}{[L(\omega)+\alpha M(\omega)]\,[L(\omega')+\alpha M(\omega')]} \times$$

$$\times \exp[-i(\omega+\omega')t]\, d\omega\, d\omega'$$

since $\overline{v(\omega)} = \overline{v(\omega')} = 0$. For stationary random processes,

$$\overline{z_T(\omega) \cdot z_T(\omega')} = N(\omega)\,\delta(\omega+\omega')$$

and

$$\overline{v(\omega)\,v(\omega')} = S(\omega)\,\delta(\omega+\omega'),$$

where $\delta(\omega + \omega')$ is Dirac's delta function. Performing the integration in the expression on the right with respect to ω' and using the properties of the delta function and the fact that $M(\omega)$ and $L(\omega)$ are even functions, we obtain the value of the deviation

$$\overline{[z_\alpha(t) - z_T(t)]^2} = T(\alpha M)$$

$$T(\alpha M) = \frac{1}{4\pi^2}\int_{-\infty}^{\infty} \frac{\alpha^2 M^2(\omega)\,N(\omega) + L(\omega)\,S(\omega)}{[L(\omega) + \alpha M(\omega)]^2}\,d\omega. \quad (5.1.2)$$

From the condition that this functional be minimized on the set of functions $M(\omega)$, we find by elementary calculations that the minimum is attained with the function

$$M(\omega) = M_0(\omega) = \frac{1}{\alpha}\frac{S(\omega)}{N(\omega)}$$

and

$$\min_{\{M(\omega)\}} \overline{[z_\alpha(t) - z_T(t)]^2} = \frac{1}{4\pi^2}\int_{-\infty}^{\infty} \frac{S^2(\omega) + L(\omega)\,N(\omega)\,S(\omega)}{N(\omega)\,[L(\omega) + \alpha M_0(\omega)]^2}\,d\omega.$$

Thus, an operator $R_M(u, \alpha)$ of the form (5.1.1) minimizing the deviation (5.0.2) of the regularized solution $z_\alpha(t)$ from the exact one $z_T(t)$ and corresponding to the function

$$M(\omega) = M_0(\omega) = \frac{1}{\alpha}\frac{S(\omega)}{N(\omega)}$$

is independent of α and has the form

$$R_{M_0}(u) = \frac{1}{2\pi} \int_{-\infty}^{\infty} \frac{K(-\omega)\, u(\omega)}{L(\omega) + \frac{S(\omega)}{N(\omega)}} \exp(-i\omega t)\, d\omega. \quad (5.1.3)$$

The approximate (regularized) solution $z_{op}(t)$ of equation (5.0.1) obtained with the aid of this operator is independent of the parameter α and it is represented by the formula

$$z_{op}(t) = \frac{1}{2\pi} \int_{-\infty}^{\infty} \frac{K(-\omega)\, u(\omega)}{L(\omega) + \frac{S(\omega)}{N(\omega)}} \exp(-i\omega t)\, d\omega. \quad (5.1.4)$$

We shall call it the **optimal regularized solution** of equation (5.0.1). It coincides with the result of applying the optimal Wiener filtering to find $z(t)$ from $u(t) = u_T(t) + v(t)$. [220]. Therefore, we shall also call the operator $R_{M_0}(u)$ the **optimal Wiener filtering operator**.

Remark. To obtain the optimal regularized solution, we need to know the spectral densities of the noise and the solution that we are seeking. In practical problems leading to equation (5.0.1), we often have information enabling us to find $S(\omega)$ (at least approximately). We usually make some sort of assumptions regarding the spectral density of the solution $N(\omega)$ and then check to see if they are valid. In §3 of the present chapter, we shall give a method for determining approximately the asymptotic behavior of the functions $S(\omega)$ and $N(\omega)$ as $\omega \to \infty$ under the supplementary assumption that the random processes in question are ergodic.

3. Formula (5.1.1) defines one-parameter families (classes) of linear regularizing operators. Each such family is defined in terms of the function $M(\omega)$ satisfying conditions a)–d) of Chapter IV, §1. Thus, we have proven

Theorem 1. Among the functions $M(\omega)$ possessing properties a)–d) of Chapter IV, §1, there exists a function $M_0(\omega)$ such that the one-parameter family of regularizing operators $R_{M_0}(u, \alpha)$ includes the optimal Wiener filtering operator.

The significance of Theorem 1 lies, first of all, in the fact that the optimal filtering operator belongs to the family of stable operators that depend continuously on the right-hand member $u(t)$ of equation (5.0.1). Consequently, the optimal filtering operator is stable, that is, continuous with respect to u at every fixed exact right-hand member. Second, it follows from this theorem and §3 of Chapter IV that small deviations of the function $M(\omega)$ from $S(\omega)/N(\omega)$ (in the sense of asymptotic ϵ-closeness of the corresponding factors $f(\omega, \alpha)$) cause small deviations in the regularized solution from the optimal one. This reveals the possibility of obtaining approximate solutions of equation (5.0.1) that are close to the optimal one by using simpler algorithms for constructing the regularized solution.

In many cases, it is possible to calculate (most often approximately) the spectral density of the noise $S(\omega)$ by using information regarding the noise but the spectral density of the solution $N(\omega)$ that we are seeking is unknown. Therefore, it is usually impossible to use the optimal filtration operator to construct an approximate solution. In this connection, the possibility of replacing the optimal regularizing operator (the optimal filtering operator) with an operator close to it, one using less *a priori* information for its construction, is of interest. In §3 of the present chapter, we shall, for equations of types I and II, demonstrate the possibility of finding, to any desired degree of accuracy, the asymptotic behaviors of $S(\omega)$ and $N(\omega)$ from the family of regularized solutions under the assumption that the stationary random processes in question are ergodic. Here, we shall use the simplest pth-order regularizing operators ($M(\omega) = \omega^{2p}$).

4. If $M(\omega) = \omega^{2p}$, where p is a nonnegative number, not necessarily an integer, then $T(\alpha M)$ is a function of the parameters α and p. Let us denote it by $T(\alpha, p)$.

If we take $M(\omega) = \omega^{2p}$ in formula (5.1.2), we obtain

$$T(\alpha, p) = \frac{1}{2\pi^2}\int\limits_0^\infty \frac{L \cdot S \cdot d\omega}{(L+\alpha\omega^{2p})^2} + \frac{1}{2\pi^2}\int\limits_0^\infty \frac{\alpha^2\omega^{4p}N \cdot d\omega}{(L+\alpha\omega^{2p})^2}. \quad (5.1.5)$$

For any $p > 0$, when we let α approach zero, the second integral in (5.1.5), namely,

$$\Delta^2(\alpha) = \frac{1}{2\pi^2}\int\limits_0^\infty \frac{\alpha^2\omega^{4p} \cdot N \cdot d\omega}{(L+\alpha\omega^{2p})^2},$$

approaches zero monotonically* and the integral

$$\sigma^2(\alpha) = \frac{1}{2\pi^2}\int\limits_0^\infty \frac{L \cdot S \cdot d\omega}{(L+\alpha\omega^{2p})^2}$$

decreases monotonically.

Furthermore, $\Delta^2(\infty) = \infty$ and $\sigma^2(\infty) = 0$. Therefore, for every fixed value of p, the function

$$T(\alpha, p) = \sigma^2(\alpha) + \Delta^2(\alpha)$$

has a mininıum at some** value $\alpha = \alpha_p$. We shall call this value the p-optimal value of the regularization parameter α. It is determined from the condtion $\partial T/\partial\alpha = 0$, which has the form

$$\int\limits_0^\infty \frac{\alpha \cdot L \cdot N \cdot \omega^{4p}}{(L+\alpha\omega^{2p})^3} d\omega = \int\limits_0^\infty \frac{S \cdot L \cdot \omega^{2p}\, d\omega}{(L+\alpha\omega^{2p})^3}. \quad (5.1.6)$$

Obviously, α_p is a function of p. Consequently,

$$T(\alpha_p, p) = \psi(p). \quad (5.1.7)$$

---

*This can be proven just as for $\Delta_r(t, \alpha)$ in Chapter IV.

**If this value is not unique, let us take the smallest possible value.

5. Let us examine some of the properties of the function $\psi(p)$. We shall do this for equations with kernels $K(t)$ whose Fourier transforms on the real axis have an asymptotic behavior, as $\omega \to \infty$, of one of the following types:

1) $K(\omega) = \dfrac{A}{\omega^n}$, $n > 0$ (type I);

2) $K(\omega) \doteq H\omega^\nu \exp[-(i\omega A)^{\frac{1}{m}}]$, $A > 0$, $\nu \geqslant 0$, $H > 0$, $m > 0$ (type II);

3) $K(\omega) = H \exp(-A\omega^2)$, $A > 0$, $H > 0$ (type III);

4) $K(\omega) = \dfrac{A}{\omega^n (\underbrace{\ln \ln \ldots \ln}_{s \text{ times}} \omega)^k}$, $n \geqslant 0$, $k \geqslant 0$, $n + k \neq 0$ (type IV).

We shall also assume that the spectral densities of $N(\omega)$ and $S(\omega)$ have asymptotic representations as $\omega \to \infty$ *on the real axis* of the form

$$N(\omega) = \frac{N_0}{\omega^{2b}}, \quad b > 0; \quad S(\omega) = \frac{S_0}{\omega^{2a}}, \quad a \geqslant 0.$$

Let us look at asymptotic estimates of the integrals

$$\Delta^2(\alpha) = \frac{1}{2\pi^2} \int_0^\infty \frac{\alpha^2 \omega^{4p} N \, d\omega}{(L + \alpha\omega^{2p})^2} \quad \text{and} \quad \sigma^2(\alpha) = \frac{1}{2\pi^2} \int_0^\infty \frac{S \cdot L \, d\omega}{(L + \alpha\omega^{2p})^2}.$$

§2. Properties of the function $\psi(p)$ for equations with kernels of types I–IV.

1. Kernels of type I. Repeating the reasoning of Chapter IV, §3, we find

$$\Delta^2(\alpha) = \frac{1}{2\pi^2}\int_0^\infty \frac{\alpha^2\omega^{4p}N_{as}\,d\omega}{(L_{as}+\alpha\omega^{2p})^2} + O(\alpha^2),$$

$$\sigma^2(\alpha) = \frac{1}{2\pi^2}\int_0^\infty \frac{S_{as}L_{as}d\omega}{(L_{as}+\alpha\omega^{2p})^2} + O(S_0),$$

where, as before, the symbol f_{as} means that $f(\omega)$ is calculated for all values of ω from the asymptotic formula. Thus,

$$\frac{1}{2\pi^2}\int_0^\infty \frac{\alpha^2\omega^{4p}N_{as}}{(L_{as}+\alpha\omega^{2p})^2} =$$

$$= \frac{N_0}{2\pi^2|A^4|}\int_0^\infty \frac{\alpha^2\omega^{4p-2b}\,d\omega}{\left(1+\frac{\alpha}{|A|^2}\omega^{2q}\right)^2}, \quad q = p+n.$$

This integral can be evaluated* by making the change of variable $x = (\alpha/|A|^2)\omega^{2q}$. Proceeding in this way, we obtain

$$\Delta^2(\alpha) = \frac{N_0}{4\pi q}(1-\gamma)\frac{1}{\sin\gamma\pi}\left(\frac{\alpha}{|A|^2}\right)^\gamma + O(\alpha^2), \qquad (5.2.1)$$

where $\gamma = (2b-1)/2q$.

In an analogous manner, we calculate $\sigma^2(\alpha)$:

$$\sigma^2(\alpha) = \frac{S_0}{4\pi q|A|^2}(1-\mu)\frac{1}{\sin\mu\pi}\left(\frac{\alpha}{|A|^2}\right)^{-\mu} + O(S_0), \qquad (5.2.2)$$

---

*I. S. Gradshteyn and I. M. Ryzhik, *Table of Integrals, Series and Products*, New York and London, Academic Press, 1965, 4th ed. (translation of *Tablitsy integralov, summ, ryadov i proizvedeniy*), p. 293.

where

$$\mu = \frac{2n - 2a + 1}{2q} .$$

Consequently, for type I kernels, we have

$$T(\alpha, p) = C_1 \left(\frac{\alpha}{|A|^2} \right)^{\gamma} + C_2 \left(\frac{\alpha}{|A|^2} \right)^{-\mu} S_0 + O(\alpha^2) + O(S_0) \tag{5.2.3}$$

as $\alpha \to 0$, where

$$C_1 = \frac{N_0}{4\pi q} \frac{1-\gamma}{\sin \gamma\pi}; \quad C_2 = \frac{1}{4\pi q |A|^2} \frac{1-\mu}{\sin \mu\pi} .$$

Let us suppose that nonnegative numbers p, n, b, and a satisfy the conditions

$$0 < \gamma < 1, \quad 0 < \mu < 1, \text{ and } b \geqslant 2a, \tag{5.2.4}$$

from which it follows that

1) $2b > 1$; $\qquad 2b < 2n + 2p + 1$;
2) $2a < 2n + 1$; $2p > 1 - 2a$.

It follows from these requirements, in particular, that, in the case of a white noise ($a = 0$), we need to use a regularizing algorithm of order $p > ½$.

The numbers b and n characterize the order of smoothness of the solution sought and of the kernel $K(t)$. The inequality $2b < 2n + 2p + 1$ indicates that, when $n < b$, the order p of the regularization cannot be chosen arbitrarily. We mention that the numbers p, n, b, and a are not necessarily integers.

Using formula (5.2.3) and keeping only the principal terms, we find from the condition $\partial T/\partial\alpha = 0$ the p-optimal value* of α:

$$\alpha_p = |A|^2 \left(\frac{\mu}{\gamma}\frac{C_2}{C_1} S_0\right)^{\frac{1}{\gamma+\mu}} \tag{5.2.5}$$

or

$$\alpha_p = |A|^2 \left(\frac{S_0}{N_0 |A|^2}\frac{\mu}{\gamma}\frac{1-\mu}{1-\gamma}\frac{\sin\gamma\pi}{\sin\mu\pi}\right)^{\frac{1}{\gamma+\mu}}. \tag{5.2.6}$$

Substituting this value of α_p into formula (5.2.3), we obtain

$$\psi(p) = T(\alpha_p, p) = \left(\frac{\mu}{\gamma}\frac{C_2}{C_1} S_0\right)^{\frac{\gamma}{\gamma+\mu}} \cdot \left(1+\frac{\gamma}{\mu}\right) C_1 \tag{5.2.7}$$

or

$$\psi(p) = \left(1+\frac{\gamma}{\mu}\right)\frac{N_0}{4q\pi}\frac{1-\gamma}{\sin\gamma\pi}\left\{\frac{\mu(1-\mu)}{\gamma(1-\gamma)}\frac{\sin\gamma\pi}{\sin\mu\pi}\frac{S_0}{N_0|A|^2}\right\}^{\frac{\gamma}{\gamma+\mu}}. \tag{5.2.8}$$

Of course, γ and μ are functions of p. Here, $\psi(p)$ is the *smallest* of the possible mean values (for different values of the regularization parameter α) of the square of the deviation of the regularized solution $z_\alpha(t)$ from the exact solution $z_T(t)$ in the use of the simplest pth-order regularization. How will this minimum deviation vary with change in p? The answer is given by

---

*We can obtain the same value of α from equation (5.1.6) by evaluating the integrals in it with the same change of variable $x = (\alpha/|A|^2)\omega^{2q}$.

Theorem 2. For type-I kernels, the function $\psi(p)$ has the following properties:

1) it has a unique minimum at $p = p_0 = b - a$;

2) it increases monotonically in the interval $p > p_0$ and, as $p \to \infty$, it approaches a finite limit

$$\psi(\infty) = \psi(p_0)\frac{\sin(\gamma_0\pi)}{\pi\gamma_0(1-\gamma_0)}; \tag{5.2.9}$$

3) $$\frac{\psi''(p_0)}{\psi(p_0)} = \frac{y(1-y)}{(2b-1)^2}\left\{\frac{\pi^2y^2}{\sin^2(\pi y)} - 1 - \frac{y^2}{(1-y)^2}\right\}, \tag{5.2.10}$$

where

$$y = \frac{\gamma_0}{\mu_0}, \quad \gamma_0 = \gamma(p_0), \quad \mu_0 = \mu(p_0).$$

Proof. Taking the logarithmic derivative of $\psi(p)$, we find

$$\psi'(p) = \psi(p)\frac{\gamma\mu}{q(\gamma+\mu)}\cdot\varphi(\gamma,\mu),$$

where

$$\varphi(\gamma,\mu) = \frac{1}{1-\gamma} - \frac{1}{\gamma} + \frac{1}{1-\mu} - \frac{1}{\mu} + \pi\frac{\sin[\pi(\gamma+\mu)]}{\sin(\gamma\pi)\cdot\sin(\mu\pi)}.$$

Since $0 < \gamma < 1$ and $0 < \mu < 1$, the function $\psi(p)$ is positive for all values of p. Under the restrictions on γ and μ mentioned, the function $\varphi(\gamma, \mu)$ vanishes only when $\gamma(p) + \mu(p) = 1$, that is, when $p = p_0 = b - a$. This proves property 1).

The function $\varphi(\gamma, \mu)$ remains positive in the interval $p > p_0$. Consequently, in this interval, $\psi(p)$ increases monotonically.

Using formula (5.2.8) and taking the limit directly, we find

$$\psi(\infty) = \frac{1}{\mu_0}\frac{N_0}{2\pi^2(2b-1)}\left\{\frac{S_0}{N_0|A|^2}\right\}^{\gamma_0}.$$

On the other hand,

$$\psi(p_0) = \frac{\gamma_0(1-\gamma_0)}{2\pi \sin(\pi\gamma_0)} \left\{\frac{S_0}{N_0 |A|^2}\right\}^{\gamma_0} \frac{N_0}{\mu_0(2b-1)}.$$

Equation (5.2.9) follows from these formulas. Property 3) is established by a straightforward calculation beginning with formula (5.2.8),

Corollary 1. *For arbitrary* $p > p_0$,

$$0 < \frac{\psi(p) - \psi(p_0)}{\psi(p_0)} < \frac{\gamma_0}{1-\gamma_0}.$$

Thus, the smallest square deviation $\psi(p)$, corresponding to regularization of arbitrary order $p > p_0$, differs only slightly from the smallest square deviation $\psi(p_0)$ corresponding to regularization *of optimal order* p_0. Consequently, *for sufficiently small values of* γ_0, *we can use a regularization of arbitrary sufficiently large order* $p > p_0$. We note that we then need to take the regularization parameter α equal to α_p.

We shall say that the operator

$$Az \equiv \int_{-\infty}^{\infty} K(t-\tau) z(\tau) \, d\tau$$

is **strongly smoothing** [10, 11] if n is considerably greater than b and a, that is, if $n \gg \max\{b, a\}$.

Corollary 2. *For a strongly smoothing operator Az, the minimum of the mean value of the square deviation of the regularized solution from the exact one, that is, the function* $\psi(p)$, *is weakly dependent on the order p of the regularization.*

This is true because, in such a case, the number

$$\gamma_0 = \frac{2b-1}{2n+2b-2a}$$

is small.

We shall call the simplest stabilizer of order p_0 the **best simplest stabilizer**. Consequently, for a strongly smoothing operator Az, one can in computational work use the simplest stabilizer of any sufficiently large order $p > p_0$, not necessarily the best.

Remark 1. The ratio $\frac{\psi(p) - \psi(p_0)}{\psi(p_0)}$ can also be estimated by using property 3) of the function $\psi(p)$ in Theorem 2. We obtain

$$\frac{\psi(p) - \psi(p_0)}{\psi_0(p_0)} \approx \frac{(p - p_0)^2}{2} \frac{y(1-y)}{(2b-1)^2} \left\{ \frac{\pi^2 y^2}{\sin^2(\pi y)} - 1 - \frac{y^2}{(1-y)^2} \right\}.$$

We shall call the value $\alpha_p(p)$ of α corresponding to regularization of order $p = p_0$ the **optimal value** of the regularization parameter and we shall denote it by α_{opt}:

$$\alpha_{\text{opt}} = \alpha_p(p_0).$$

Obviously, for equations with type I kernels,

$$\alpha_{\text{opt}} = S_0/N_0.$$

The regularized solution $z_{\text{opt}}(t)$ obtained by regularization of order $p = p_0$ with $\alpha = \alpha_{\text{opt}}$ coincides with the function $z_{\text{o.f.}}(t)$ obtained by optimal Wiener filtering; that is,

$$z_{\text{opt}}(t) \equiv z_{\text{o.f.}}(t).$$

Remark 2. It follows from formulas (5.2.1) and (5.2.2) that, for $p = p_0$ and $\alpha = \alpha_{\text{opt}}$,

$$\frac{\sigma^2(\alpha_{\text{opt}})}{\Delta^2(\alpha_{\text{opt}})} = \frac{2b-1}{2n-2a+1} = y.$$

Consequently, *for strongly smoothing operators* ($n \gg a, b$), when one is using regularization of optimal order p_0 with the optimal value of the regularization parameter ($\alpha = \alpha_{\text{opt}}$), the chief contribution to the size of the deviation $T(\alpha_{\text{opt}}, p_0)$ *is that of the regularization*, not that of the noise.

2. Kernels of types II and III. By virtue of what was shown at the beginning of subsection 1, we can replace equation (5.1.6) for determining α_p with the equation

$$\alpha N_0 \int_0^\infty \frac{\omega^{4p-2b} L_{\text{as}}(\omega)\, d\omega}{\{L_{\text{as}}(\omega) + \alpha\omega^{2p}\}^3} = S_0 \int_0^\infty \frac{\omega^{2p-2a} L_{\text{as}}(\omega)\, d\omega}{\{L_{\text{as}}(\omega) + \alpha\omega^{2p}\}^3}. \qquad (5.2.11)$$

The integral

$$I_1 = \int_0^\infty \frac{\omega^{2p-2a} L_{\text{as}}}{(L_{\text{as}} + \alpha\omega^{2p})^3}\, d\omega$$

is evaluated by the method of steepest descent and is equal to

$$I_1 = \frac{\omega_1^{2p-2a} L_{\text{as}}(\omega_1)}{\{L_{\text{as}}(\omega_1) + \alpha\omega_1^{2p}\}^3} \left[\sqrt{\frac{\pi}{|f_1''(\omega_1, \alpha)|}} + O(\alpha^{\frac{3}{2}}) \right],$$

where

$$f_1(\omega, \alpha) = \alpha \ln \frac{\omega^{2p-2a} L_{\text{as}}(\omega)}{\{L_{\text{as}}(\omega) + \alpha\omega^{2p}\}^3}$$

and the double prime denotes the second derivative with respect to ω. Here, ω_1 is a root of the equation

$$(-2L_{\text{as}} + \alpha\omega^{2p}) \frac{dL_{\text{as}}}{d\omega} - \frac{4p+2a}{\omega} \alpha \cdot \omega^{2p} \cdot L_{\text{as}} + \frac{2p-2a}{\omega} L_{\text{as}}^2 = 0.$$

Obviously, ω_1 is a function of α and it approaches ∞ as $\alpha \to 0$. Since

$$\frac{\omega^{4p-2b} L_{\text{as}}(\omega)}{\{L_{\text{as}}(\omega) + \alpha\omega^{2p}\}^3} = \omega^{2p-2b+2a} \exp\left[\frac{1}{\alpha} f_1(\omega, \alpha)\right],$$

the integral

$$I_2 = \int_0^\infty \frac{\omega^{4p-2b} L_{as} d\omega}{(L_{as} + \alpha\omega^{2p})^3}$$

is also evaluated by the method of steepest descent and is equal to

$$I_2 = \frac{\omega_1^{2p-2a} L_{as}(\omega_1)}{\{L_{as}(\omega_1) + \alpha\omega_1^{2p}\}^3} \left[\sqrt{\frac{\pi}{|f_1''(\omega_1, \alpha)|}}\, \omega_1^{2p-2b+2a} + O\left(\alpha^{\frac{3}{2}}\right) \right].$$

Substituting the values found for I_1 and I_2 into formula (5.2.11) and keeping only the principal terms, we obtain the following equation for determining α_p:

$$\alpha = \frac{S_0}{N_0} \{\omega_1(\alpha)\}^{-2p-2b+2a}.$$

For $p = p_0$, we obtain $\alpha_{opt} = S_0/N_0$.

In subsection 1, we saw that the greater n is, that is, the higher the order of decrease in $K(\omega)$ as $\omega \to \infty$, the smaller will the difference $\psi(p) - \psi(p_0)$ be for $p > p_0$, that is, the weaker will the influence of the order of regularization p on the function $\psi(p)$ be. Consequently, we can quite justifiably replace the optimal order of regularization p_0 with an arbitrary order $p > p_0$.

In the case of equations with kernels of types II and III, the order of decrease of the function $K(\omega)$ as $\omega \to \infty$ is greater than that of any power with negative exponent ω^{-n}. Therefore, we can regard an operator Az with kernel of type II or III as strongly smoothing. Consequently, Corollary 2 of Theorem 2 of the present section is applicable to it.

In the case of an equation with type-IV kernel, everything is determined by the value of the exponent n rather than by the numbers s and k. We shall not carry out the corresponding calcuations [13].

Remark 3. Results analogous to those of §§ 1 and 2 remain valid for multidimensional integral equations of the first kind of the convolution type (see [144, 145]).

§3. Determination of the high-frequency characteristics of a signal and noise and of the optimal value of the regularization parameter.

1. Let us return to the question of finding the optimal regularization order p_0 and the optimal value of the regularization parameter α_{opt}. As was shown for type-I equations, $p_0 = b - a$ and $\alpha_{\text{opt}} = S_0/N_0$. Thus, to find p_0 and α_{opt} (and also the p-optimal value α_p of α), we need to know the high-frequency characteristics of the signal and noise, namely, N_0, b, S_0, and a. In many practical problems, the characteristics of the noise S_0 and a can be determined (approximately) from observational results. This is not true of the high-frequency characteristics of the signal N_0 and b. Our assumption regarding the nature of the dependence of the spectral density of the signal on the frequency, that is, $N(\omega)$, is valid for a broad class of problems. However, the numbers N_0 and b, which determine this dependence for large values of ω, are usually unknown and it is not possible to determine them directly from observational results.

2. In the present section, we shall show that these parameters can be uniquely determined from the family of regularized solutions $\{z_\alpha(t)\}$ corresponding to different values of α and constructed with the aid of the simplest regularizing operator $M(\omega) = \omega^{2p}$, the order p of which can be chosen rather arbitrarily. This can be done in an especially simple manner for *ergodic* processes [11].

Thus, application of the regularization method to ergodic processes enables us to find the optimal approximate solution of equation (5.0.1) (or one close to it), using much less *a priori* information regarding the solution sought and the noise than with the method of optimal filtering.

3. Let us define

$$F(\alpha) = \alpha^{\frac{-2n}{n+p}} \int_{-\infty}^{\infty} \overline{\{A^* [u(t) - Az_\alpha(t)]\}^2 \, dt},$$

where A^* is the adjoint of the operator A in L_2. Obviously, $F(\alpha) > 0$ for all $\alpha > 0$.

Let us look at equations with type-I kernels.

Theorem. *For small noises (that is, for small S_0), the function $F(\alpha)$ has a local minimum at a point $\alpha = \alpha_{min}$ and a local maximum at a point $\alpha = \alpha_{max}$. Here, $\alpha_{max} < \alpha_{min}$.*

Proof: Using Plancherel's theorem and the expression for the spectral density $U(\omega)$ of the right-hand member of equation (5.0.1)

$$U(\omega) = L(\omega) N(\omega) + S(\omega),$$

we find

$$F(\alpha) = \alpha^{\frac{2p}{q}} \int_{-\infty}^{\infty} \frac{L(LN + S)\,\omega^{4p}\, d\omega}{(L + \alpha\omega^{2p})^2}$$

or

$$F(\alpha) = 2\alpha^{\frac{2p}{q}} \left\{ \int_0^{\omega_0} \frac{L(LN + S)\,\omega^{4p}}{(L + \alpha\omega^{2p})^2}\, d\omega + \int_{\omega_0}^{\infty} \frac{L(LN + S)\,\omega^{4p}}{(L + \alpha\omega^{2p})^2}\, d\omega \right\}.$$

We choose ω_0 so that, for $\omega > \omega_0$, the function $L(\omega)$, $N(\omega)$, and $S(\omega)$ can be calculated from asymptotic formulas.

For small α,

$$\int_0^{\omega_0} \frac{L^2 N \omega^{4p}\, d\omega}{(L + \alpha\omega^{2p})^2} = B_1 \int_0^{\omega_0} N\omega^{4p}\, d\omega = O(1), \text{ where } 0.25 < B_1 < 1;$$

$$\int^{\omega_0} \frac{LS\omega^{4p}\, d\omega}{(L + \alpha\omega^{2p})^2} = B_2 \int_0^{\omega_0} S \cdot L^{-1}\omega^{4p}\, d\omega = O(1), \text{where } 0.25 < B_2 < 1;$$

$$I_1 = \int_{\omega_0}^{\infty} \frac{L^2 N \omega^{4p} \, d\omega}{(L + \alpha\omega^{2p})^2} = N_0 \int_{\omega_0}^{\infty} \frac{\omega^{4p-2b} \, d\omega}{\left(1 + \frac{\alpha}{|A|^2} \omega^{2q}\right)^2} =$$

$$= \frac{N_0}{2q} \int_{x_0}^{\infty} \frac{x^{\frac{4p-2b+1}{2q} - 1} \, dx}{(1+x)^2} \cdot \left(\frac{\alpha}{|A|^2}\right)^{\frac{-4p+2b-1}{2q}},$$

where

$$x_0 = \frac{\alpha}{|A|^2} \omega_0^{2q}.$$

As $\alpha \to 0$, the last integral approaches the value of the same integral with lower end-point 0, which is easily evaluated and is found to be equal to

$$\pi \frac{2p - 2n - 2b + 1}{2q} \frac{1}{\sin\left(\frac{4n + 2b - 1}{2q} \pi\right)}.$$

Thus,

$$I_1 = C_1(\alpha) \left(\frac{\alpha}{|A|^2}\right)^{\frac{-4p+2b-1}{2q}},$$

where

$$C_1(\alpha) \to C_{10} = \frac{\pi N_0}{(2q)^2} \frac{2p - 2n - 2b + 1}{\sin\left(\frac{4n + 2b - 1}{2q} \pi\right)} \quad \text{as} \quad \alpha \to 0.$$

In an analogous manner, we find

$$I_2 = \int_{\omega_0}^{\infty} \frac{LS\omega^{4p}d\omega}{(L+\alpha\omega^{2p})^2} = C_2(\alpha)\left(\frac{\alpha}{|A|^2}\right)^{\frac{-4p+2n-2a+1}{2q}} \cdot \frac{S_0}{|A|^2},$$

where

$$C_2(\alpha) \to C_{20} = \frac{\pi}{(2q)^2} \frac{2p-2a+1}{\sin\left(\frac{2p-2a+1}{2q}\pi\right)} \quad \text{as} \quad \alpha \to 0.$$

Consequently, for small α,

$$F(\alpha) = C_3(\alpha)\left(\frac{\alpha}{|A|^2}\right)^{\gamma} + C_4(\alpha)\left(\frac{\alpha}{|A|^2}\right)^{-\mu} \cdot \frac{S_0}{|A^2|} + O(\alpha^{\frac{2p}{q}}), \tag{5.3.1}$$

where

$$C_3(\alpha) = 2C_1(\alpha)|A|^{\frac{-4p}{q}}, \quad C_4(\alpha) = 2C_2(\alpha)|A|^{\frac{-4p}{q}}.$$

Since γ and μ are positive, it follows that $F(\alpha) \to \infty$ as $\alpha \to 0$.

For $\alpha \gg 1$, the integral

$$\int_0^{\omega_0} \frac{\omega^{4p}L \cdot U\, d\omega}{(L+\alpha\omega^{2p})^2}$$

lies between the integrals

$$\int_0^{\omega_0} \frac{(LU)_{max}\omega^{4p}d\omega}{(L_{min}+\alpha\omega^{2p})^2} \quad \text{and} \quad \int_0^{\omega_0} \frac{(LU)_{min}\omega^{4p}d\omega}{(L_{max}+\alpha\omega^{2p})^2}.$$

But

$$\int_0^{\omega_0} \frac{\omega^{4p}d\omega}{(\beta+\alpha\omega^{2p})^2} = \frac{1}{2p}\cdot\alpha^{-2-\frac{1}{2p}} \int_0^{\alpha\omega_0^{2p}} \frac{x^{1+\frac{1}{2p}}dx}{(\beta+x)^2}, \qquad (\beta \geqslant 0).$$

As $\alpha \to \infty$, the last integral approaches infinity at the same rate as $\alpha^{1/2p}$. Consequently,

$$\int_0^{\omega_0} \frac{\omega^{4p}d\omega}{(\beta+\alpha\omega^{2p})^2} = O(\alpha^{-2})$$

and

$$\alpha^{\frac{2p}{q}} \int_0^{\omega_0} \frac{\omega^{4p}\cdot L\cdot U\, d\omega}{(L+\alpha\omega^{2p})^2} = O(\alpha^{-\frac{2n}{q}}).$$

Let us look at the integrals over the interval (ω_0, ∞):

$$\int_{\omega_0}^{\infty} \frac{L^2N\,\omega^{4p}d\omega}{(L+\alpha\omega^{2p})^2} = N_0 \int_{\omega_0}^{\infty} \frac{\omega^{4p-2b}d\omega}{\left(1+\frac{\alpha}{|A|^2}\,\omega^{2q}\right)^2} =$$

$$= \frac{N_0}{2q}\,\alpha^{\frac{-4p+2b-1}{2q}} \int_{\alpha\omega_0^{2q}}^{\infty} \frac{x^{\frac{4p-2b+1}{2q}-1}dx}{(1+x)^2}.$$

As $\alpha \to \infty$, the last integral approaches zero at the same rate as $\alpha^{\frac{-4n-2b+1}{2q}}$. Consequently,

$$\int_{\omega_0}^{\infty} \frac{L^2N\omega^{4p}d\omega}{(L+\alpha\omega^{2p})^2} = O(\alpha^{-2})$$

and

$$\alpha^{\frac{2p}{q}} \int_{\omega_0}^{\infty} \frac{L^2 N\omega^{4p} d\omega}{(L + \alpha\omega^{2p})^2} = O(\alpha^{\frac{-2n}{q}}).$$

In an analogous manner, we find

$$\alpha^{\frac{2p}{q}} \int_{\omega_0}^{\infty} \frac{LS\,\omega^{4p} d\omega}{(L + \alpha\omega^{2p})^2} = O(\alpha^{\frac{-2n}{q}}).$$

Thus, as $\alpha \to \infty$,

$$F(\alpha) = O(\alpha^{\frac{-2n}{q}}).$$

Let us estimate the derivative $F'(\alpha)$ for small values of α. Obviously,

$$\frac{d}{d\alpha} [\ln F(\alpha)] = \frac{2p}{\alpha q} + \frac{d}{d\alpha} [\ln F_1(\alpha)],$$

where

$$F_1(\alpha) = \int_0^{\infty} \frac{L(LN + S)\,\omega^{4p} d\omega}{(L + \alpha\omega^{2p})^2}.$$

The condition $F'(\alpha) = 0$ is equivalent to the condition

$$\alpha F_1'(\alpha) + \frac{2p}{q} F_1(\alpha) = 0$$

or

$$\alpha^{\frac{2p}{q}+1} F_1'(\alpha) + \frac{2p}{q} F(\alpha) = 0. \qquad (5.3.2)$$

Repeating the reasoning followed in the estimation of $F(\alpha)$ for large values of α, we find that, for small values of α, the left-hand member of equation (5.3.1) has the form

$$\widetilde{C}_1(\alpha)\gamma\left(\frac{\alpha}{|A|^2}\right)^{\gamma}-\widetilde{C}_2(\alpha)\mu\left(\frac{\alpha}{|A|^2}\right)^{-\mu}\frac{S_0}{|A|^2}+O\left(\alpha^{\frac{2p}{q}}\right), \qquad (5.3.3)$$

where

$$\widetilde{C}_1(\alpha)\to 2|A|^{-4p/q}C_{10},\ \widetilde{C}_2(\alpha)\to 2|A|^{-4p/q}C_{20} \quad \text{as} \quad \alpha\to 0.$$

We can write the sum (5.3.3) in the form

$$\widetilde{C}_1(\alpha)\gamma\alpha_1^{-\mu}\left\{\alpha_1^{\gamma+\mu}+\frac{\mu}{\gamma}\frac{\widetilde{C}_2(\alpha)}{\widetilde{C}_1(\alpha)}\frac{S_0}{|A|^2}+O\left(\alpha_1^{\frac{2p}{q}+\mu}\right)\right\},\ \alpha_1=\frac{\alpha}{|A|^2}.$$

Since the numbers γ and μ, the sum $2p/q+\mu$, and the ratio $\widetilde{C}_2/\widetilde{C}_1$ are all positive, it follows that, for small values of S_0, the function

$$f_2(\alpha_1)=\alpha_1^{\gamma+\mu}-\frac{\mu}{\gamma}\frac{S_0}{|A|^2}\frac{\widetilde{C}_2}{\widetilde{C}_1}+O\left(\alpha_1^{\frac{2p}{q}+\mu}\right)$$

vanishes at a point $\alpha_1=\alpha_{1,\min}$ such that $\alpha_{\min}=\alpha_{1,\min}\cdot|A|^2$ lies in the region of applicability of the representation (5.3.1) for the function $F(\alpha)$ (see Fig. 8). The point $\alpha=\alpha_{\min}$ can only be a local minimum of the function $F(\alpha)$. Obviously,

$$\alpha_{\min}=\alpha_{\min}(S_0)$$

and

$$\lim_{S_0\to 0}\alpha_{\min}(S_0)=0.$$

From the estimate of the function $F(\alpha)$ for large α, we see that,

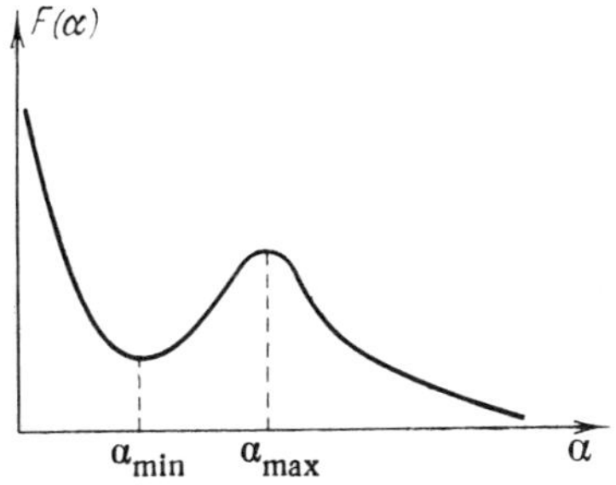

FIG. 8.

in an interval of values of α much greater than 1, the function $F(\alpha)$ approaches 0 monotonically as $\alpha \to \infty$. This fact and the fact that $\alpha = \alpha_{\min}$ $F(\alpha)$ has a local minimum imply the existence of a point $\alpha = \alpha_{\max} > \alpha_{\min}$ at which $F(\alpha)$ has a local maximum. Obviously, $\alpha_{\max}$ is a function of S_0. This completes the proof of the theorem.

4. It should be pointed out that the ratio $\alpha_{\min}/\alpha_{\max}$ approaches 0 as $S_0 \to 0$. The condition

$$\frac{\alpha_{\min}}{\alpha_{\max}} \ll 1$$

may be regarded as a criterion of smallness of the noise and hence as a criterion of applicability of the above-obtained results to the analysis of experimental curves $u(t)$.

For values of p satisfying the inequality

$$4p > 2b - 1,$$

the first term in the formula for $f_2(\alpha_1)$ can be neglected. In this case, we find from the condition $f_2(\alpha_1) = 0$

$$\alpha_{\min} = \left(\frac{\widetilde{C}_2}{\widetilde{C}_1} \frac{\mu}{\gamma} \frac{S_0}{|A|^2} \right)^{\frac{1}{\gamma+\mu}}.$$

In the case of *ergodic* processes, the function $F(\alpha)$ is determined without *a priori* knowledge of the high-frequency characteristics of the solution sought and the noise from the formula

$$F(\alpha) = \alpha^{\frac{-2n}{q}} \lim_{\to\infty} \frac{1}{T} \int_0^T \{A^* (Az_\alpha - u)\}^2 \, dt.$$

We can then determine unambiguously the values of the parameters N_0, b, S_0, and a.

Specifically, on the interval $(0, \alpha_{min})$ the curve $y = F(\alpha)$ is determined only by the parameters a, S_0, p, and n since the term

$$C_4(\alpha) \frac{S_0}{|A|^2} \left(\frac{\alpha}{|A|^2}\right)^{-\mu}$$

is the predominant term in the formula for $F(\alpha)$ for small values of α, where

$$C_4(\alpha) = 2C_2(\alpha) \, |A|^{\frac{-4p}{q}}$$

and

$$C_2(\alpha) \to \frac{\pi}{(2q)^2} \frac{2p - 2a + 1}{\sin\left(\dfrac{2p - 2a + 1}{2q} \pi\right)} \quad \text{as} \quad \alpha \to 0.$$

On the interval $(\alpha_{min}, \alpha_{max})$, the curve $y = F(\alpha)$ is determined only by the parameters N_0, b, p, and n since on this interval the predominant term in the formula for $F(\alpha)$ is

$$C_3(\alpha) \left(\frac{\alpha}{|A|^2}\right)^{\gamma},$$

where

$$C_3(\alpha) = 2 \, |A|^{\frac{-4p}{q}} \, C_1(\alpha)$$

and

$$C_1(\alpha) \rightarrow \frac{\pi N_0}{(2q)^2} \frac{2p - 2n - 2b + 1}{\sin\left(\frac{2p - 2n - 2b + 1}{2q}\pi\right)} \quad \text{as} \quad \alpha \rightarrow 0.$$

Since, for ergodic processes, $F(\alpha)$ is determined without *a priori* information regarding the high-frequency characteristics of the signal and noise, we can in these cases determine the parameters S_0, a, N_0, and b from the regularized solutions.

For arbitrary stationary processes, these parameters are determined (for example, by the method of least squares) from the *family of regularized solutions* $\{z_\alpha(t)\}$, which contains a rather large number of functions $z_\alpha(t)$.

5. We have examined in detail the question of determination of the high-frequency characteristics of the signal and noise in the case of equations of type-I kernels. If the kernel is strongly smoothing ($n \gg b$, a), we can also take $n \gg p$. Then, the factor $\alpha^{-2n/(n+p)}$ in the function $F(\alpha)$ can be written in the form

$$\alpha^{-2\left(1-\frac{p}{n}+\frac{p^2}{n^2}-\dots\right)} \approx \alpha^{-2\left(1-\frac{p}{n}\right)},$$

where $p/n \ll 1$.

Keeping in mind both this fact and the fact that kernels of types II and III are strongly smoothing, we can determine the function $F(\alpha)$ for them:

$$F(\alpha) = \alpha^{-2+\epsilon \cdot p} \int_{-\infty}^{\infty} \overline{\{A^* [Az_\alpha - u]\}^2\, dt},$$

where $\epsilon \cdot p \ll 1$.

With the function $F(\alpha)$ determined in this way, if we repeat the reasoning followed for kernels of type I, we get the same results regarding the possibility of determining the high-frequency characteristics of the signal and noise.

Estimates of the convergence of regularized solutions are also contained in [44, 45, 130]. Conditions under which equations of the convolution type are well posed are examined in [83].

CHAPTER VI

STABLE METHODS OF SUMMING FOURIER SERIES WITH COEFFICIENTS THAT ARE APPROXIMATE IN THE l_2 METRIC

1. The problem of summing a Fourier series with respect to a given orthonormal system of functions $\{\varphi_n(t)\}$ consists in finding a function $f(t)$ from its Fourier coefficients.

In practice, one sometimes determines a function $f(t)$ characterizing a process or phenomenon being studied by measuring the coefficients a_n in its Fourier expansion with respect to an orthonormal system of functions $\{\varphi_n(t)\}$. Such measurements are always approximate. Thus, instead of the true values a_n, we obtain approximate values c_n of the Fourier coefficients. The problem arises of summing the Fourier series with approximate coefficients.

It was mentioned in the introduction (see Example 3) that the problem of summing Fourier series does not have the property of stability under small changes (in the l_2-metric) in the Fourier coefficients if the deviation of the sum of the series is estimated in the C-metric. Hence, this is an ill-posed problem.

A method of summing Fourier series with approximate coefficients that has been used for a long time consists in taking for the approximate value of the sum $f(t)$ of such a series the sum of a finite (and not too large) number of its initial terms; that is, one takes

$$f(t) \approx \sum_{n=1}^{k} c_n \varphi_n(t).$$

A summation method based on the idea of regularization, one that is stable under small changes in the coefficients in the l_2-metric, was proposed in [158].

Following [158], we shall say that a method of summing the Fourier series of functions $f(t) \in F$ with coefficients c_n that are approximate in the l_2-metric is stable in the sense of the metric of the space F if, for every positive number ϵ, there exists a positive number $\delta(\epsilon)$ such that the inequality

$$\sum_{n=1}^{\infty} (c_n - \tilde{c}_n)^2 \leqslant \delta^2(\epsilon)$$

implies the inequality

$$\rho_F(f, \tilde{f}) \leqslant \epsilon,$$

where f and $\tilde{f}$ are the results of summing the two series

$$\sum_{n=1}^{\infty} c_n \varphi_n(t) \text{ and } \sum_{n=1}^{\infty} \tilde{c}_n \varphi_n(t)$$

respectively by the method in question.

2. We denote by C_D the space of functions that are continuous (with respect to the C-metric) in a closed bounded region $\overline{D}$. We shall examine Fourier series of functions $f(t)$ in C_D with respect to a system $\{\varphi_n(t)\}$.

Following [158] and [12], we shall construct in §1 a class of stable methods of summing Fourier series based on the idea of regularization. The problem of summing the Fourier series of a function $f(t)$ can be regarded as the problem of solving an operator equation in the function $f(t)$. Specificially, if to every function in the set F we assign an element u of the space l_2, namely, the sequence $\{a_n\}$ of its Fourier coefficients a_n with respect to the system $\{\varphi_n(t)\}$ with weight $\rho(t)$ (thus, $u = \{a_n\}$), we may write

$$Af = u. \tag{6.0.1}$$

Obviously, this operator from C_D into l_2 is continuous on C_D. Consequently, the problem of summing a Fourier series, which consists in finding the function $f(t)$ from the given sequence of its Fourier coefficients $u = \{a_n\}$, reduces to solving equation (6.0.1) for $f(t)$. We know from analysis that, in the class C_D, this problem has a unique solution.

§1. Classes of stable methods of summing Fourier series.

1. If we know only approximate values of the Fourier coefficients of the function that we are seeking, that is, of the right-hand member u of equation (6.0.1), we can speak only of finding an approximate solution of the problem. Since this problem is ill-posed, it is natural to use the regularization method to find an approximate solution of it.

Following [12], we take as stabilizing functional $\Omega[f]$ a functional of the form

$$\Omega[f] = \sum_{n=1}^{\infty} f_n^2 \cdot \xi_n, \tag{6.1.1}$$

which is defined by the sequence $\{\xi_n\}$. Here, the f_n are the Fourier coefficients of the function $f(t)$ with respect to a complete orthonormal system $\{\varphi_n(t)\}$ of functions (with weight $\rho(t) > 0$) and $\{\xi_n\}$ is a sequence of positive numbers whose order of growth as $n \to \infty$ is no less than $n^{2+\epsilon}$, where $\epsilon \geqslant 0$. One can easily see that, for an arbitrary positive number d, the set of functions $f(t) \in C_D$ for which $\Omega[f] \leqslant d$ is compact in C_D. (This remains true even if the sequence $\{\xi_n\}$ has order of growth no less than $n^{1+\epsilon}$, where $\epsilon > 0$, as $n \to \infty$.)

Stabilizing functions of this kind constitute a natural generalization of the functionals $\Omega[f]$ that we used in Chapter II.

To see this, suppose that $\{\varphi_n(t)\}$ is a complete orthonormal system of eigenfunctions of a boundary-value problem of the form

$$\frac{d}{dt}[k(t)\varphi'(t)] - q(t)\varphi(t) + \lambda\varphi(t) = 0, \quad 0 \leqslant t \leqslant l,$$

$$\varphi(0) = 0 = \varphi(l),$$

where $k(t) > 0$ and $q(t) \geqslant 0$ on the interval $[0, l]$ and $\{\lambda_n\}$ is the set of eigenvalues of that problem. Then, the stabilizing functional

$$\Omega[f] = \int_0^l \{k(f')^2 + qf^2\}\, dt,$$

which we used in Chapter II, §4 for the functions $f(t)$ satisfying the conditions $f(0) = f(l) = 0$, can be written in the form $\Omega[f] = \sum_{n=1}^{\infty} f_n^2 \lambda_n$, where the f_n are the coefficients in the Fourier expansion of $f(t)$ in functions of the system $\{\varphi_n(t)\}$. Specifically, integrating the first part of the integral expression for $\Omega[f]$ by parts, we get

$$\int_0^l k(f')^2\, dt = ff'k\,\Big|_0^l - \int_0^l f\frac{d}{dt}(kf')\, dt = -\int_0^l f\frac{d}{dt}(kf')\, dt.$$

Consequently, $\Omega[f] = \int_0^l f\left\{qf - \frac{d}{dt}(kf')\right\} dt$. Substituting for the function $f(t)$ in the right-hand member of this formula its Fourier series $\sum_{n=1}^{\infty} f_n \cdot \varphi_n(t)$, we obtain

$$\Omega[f] = \int_0^l \left(\sum_{m=1}^{\infty} f_m \cdot \varphi_m(t) \right) \sum_{n=1}^{\infty} f_n \left\{ q\varphi_n(t) - \frac{d}{dt}[k \cdot \varphi_n'] \right\} dt =$$

$$= \sum_{n,m=1}^{\infty} f_n f_m \int_0^l \varphi_m(t) \left\{ q \cdot \varphi_n - \frac{d}{dt}(k\varphi_n') \right\} dt.$$

Since $q\varphi_n - \frac{d}{dt}(k\varphi_n') \equiv \lambda_n \varphi_n$ and since the functions $\varphi_n(t)$ and $\varphi_m(t)$ are orthogonal for $n \neq m$, we have

$$\Omega[f] = \sum_{\substack{n=1\\m=1}}^{\infty} f_n f_m \lambda_n \int_0^l \varphi_m \varphi_n \, dt = \sum_{n=1}^{\infty} f_n^2 \lambda_n.$$

2. Let us formulate the problem more precisely. We denote by $\mathfrak{M}$ the set of all the above-mentioned sequences $\{\xi_n\}$ corresponding to all nonnegative values of ϵ [12].

Let $\{\varphi_n(t)\}$, where $t = (t_1, t_2 \ldots, t_N)$, denote a complete orthonormal system (with weight $\rho(t) > 0$) of functions defined on a closed bounded region $\bar{D}$ of R^N (the N-dimensional Euclidean space), and let $\hat{f}(t)$ denote a continuous function on $\bar{D}$ that can be represented by its Fourier expansion in the system $\{\varphi_n(t)\}$:

$$\hat{f}(t) = \sum_{n=1}^{\infty} a_n \varphi_n(t),$$

where

$$a_n = \int_D \rho(t) \hat{f}(t) \varphi_n(t) \, dt.$$

Suppose that, instead of the Fourier coefficients a_n, we know only approximate values of them $c_n = a_n + \gamma_n$. Suppose that the errors γ_n are small (in the l_2-metric) and, in particular,

$$\sum_{n=1}^{\infty} \gamma_n^2 \leqslant \delta^2.$$

Thus, instead of the sequence $\hat{u} = \{a_n\}$, we have the sequence $u_\delta = \{c_n\}$ such that $\rho_{l_2}(\hat{u}, u_\delta) \leqslant \delta$. Consequently, instead of the Fourier series with exact coefficients

$$\sum_{n=1}^{\infty} a_n \varphi_n(t),$$

we have the series with approximate coefficients

$$\sum_{n=1}^{\infty} c_n \varphi_n(t). \tag{6.1.2}$$

Our aim is to seek, in the class C_D, a function $\tilde{f}(t)$ approximating the function $\hat{f}(t)$ for which the sequence $\{c_n\}$ of coefficients is close (in the l_2-metric) to the sequence $\{a_n\}$ of the coefficients in the Fourier expansion of $\hat{f}(t)$, that is, a function with Fourier coefficients c_n such that

$$\sum_{n=1}^{\infty} (c_n - a_n)^2 \leqslant \delta^2.$$

The approximation must be such that $\rho_C(\tilde{f}, \hat{f}) \to 0$ as $\delta \to 0$.

We cannot take for $\tilde{f}(t)$ the limit $S(t)$ of the series (6.1.2) calculated according to the rule

$$S(t) = \lim_{k \to \infty} \sum_{n=1}^{k} c_n \varphi_n(t),$$

because, as was shown in the introduction, this summation is not stable under small changes (in the l_2-metric) in the coefficients c_n.

Obviously, we need to seek the solution in the class Q_δ of functions in C_D for which

$$\rho_{l_2}(Af, u_\delta) \leqslant \delta.$$

But this class is not compact. It is too broad. We need to narrow it down. To do this, let us take some fixed functional $\Omega_1[f]$ of the form (6.1.1) described in subsection 1:

$$\Omega_1[f] = \sum_{n=1}^{\infty} f_n^{\cdot 2} \xi_n,$$

where $\{\xi_n\} \in \mathfrak{M}$.

Let F_ξ denote the set of functions in C_D on which the functional $\Omega_1[f]$ is defined and let us then write

$$F_{\delta,\xi} = Q_\delta \cap F_\xi.$$

We shall seek an approximation of $\hat{f}(t)$ on the set $F_{\delta,\xi} \subset C_D$.

For definiteness, in what follows we shall examine the one-dimensional problem. In this case, t is the coordinate of a point on the real line and the region D is a finite interval (a, b).

Since the problem of finding a function from its Fourier coefficients reduces to solving the operator equation (6.0.1), it is natural to seek an approximation of the function $\hat{f}(t)$ by the regularization method. To do this, let us look at the functional

$$M^\alpha[u_\delta, f] = \rho_{l_2}^2(Af, u_\delta) + \alpha\Omega_1[f],$$

which contains a numerical parameter α (the regularization parameter). We can also write it in the form

$$M^{\alpha}[u_{\delta}, f] = \sum_{n=1}^{\infty} (f_n - c_n)^2 + \alpha \sum_{n=1}^{\infty} f_n^2 \xi_n, \qquad (6.1.3)$$

where the f_n are the Fourier coefficients of the function $f(t)$ in the expansion with respect to the system $\{\varphi_n(t)\}$ with weight $\rho(t) > 0$, that is,

$$f_n = \int_D f(t)\,\rho(t)\,\varphi_n(t)\,dt. \qquad (6.1.4)$$

Theorems 1 and 2 of §3 of Chapter II are valid for the smoothing functional $M^{\alpha}[f, u_{\delta}]$. The proof of their validity in this case can be carried out just as in Chapter II.

Thus, there exists the function $\tilde{f}_{\alpha}(t)$ in $F_{\delta,\xi}$ minimizing the functional $M^{\alpha}[f, u_{\delta}]$ on the set $F_{\delta,\xi}$. We shall use this function as approximation of $\hat{f}(t)$.

We shall call the functional $\Omega_1[f]$ the **stabilizing functional** for the problem of stable summation of series of the form (6.1.2).

Remark 1. If we take as the system $\{\varphi_n(t)\}$ the set of eigenfunctions of the boundary-value problem

$$\operatorname{div}(k \nabla \varphi) - q^2(t)\,\varphi + \lambda\rho(t)\,\varphi = 0, \quad t \in D,$$

$$\varphi|_S = 0 \quad \left(\text{or} \quad \left.\frac{\partial \varphi}{\partial n}\right|_S = 0\right),$$

where S is the boundary of the region D in which we are seeking the solution, then the functional $\Omega_1[f]$ can be taken in the form

$$\Omega_1[f] = \int_D \{k(\nabla f)^2 + q^2 f^2\}\,dt$$

or in the equivalent form

$$\Omega_1[f] = \sum_{n=1}^{\infty} f_n^2 \lambda_n,$$

where the λ_n are the eigenvalues of the boundary-value problem mentioned and the f_n are the coefficients in the Fourier expansion of $f(t)$ with respect to the system $\{\varphi_n(t)\}$ with weight $\rho(t)$.

3. We find the coefficients in the Fourier expansion of $\tilde{f}_\alpha(t)$ with respect to the system $\{\varphi_n(t)\}$ from the condition that the partial derivatives of the functional (6.1.3) with respect to the variables f_n (for $n = 1, 2, \ldots$) vanish. We obtain

$$\tilde{f}_{\alpha,n} = \frac{c_n}{1 + \alpha\xi_n}.$$

Thus the approximation that we are seeking for the function $\hat{f}(t)$ can be written in the form

$$\tilde{f}_\alpha(t) = \sum_{n=1}^{\infty} r(n, \alpha) c_n \varphi_n(t), \tag{6.1.5}$$

where

$$r(n, \alpha) = \frac{1}{1 + \alpha\xi_n},$$

and we can calculate $\tilde{f}_\alpha(t)$ from

$$\tilde{f}_\alpha(t) = \lim_{k\to\infty} \sum_{n=1}^{k} r(n, \alpha) c_n \varphi_n(t). \tag{6.1.6}$$

4. Formulas (6.1.5) and (6.1.6) define a method of summing the series

$$\sum_{n=1}^{\infty} c_n \varphi_n(t)$$

that is stable in the sense of the C-metric under small changes (in the l_2-metric) in the coefficients c_n. In fact, the procedure described for obtaining the function $\tilde{f}_\alpha(t)$ can be written in the form of an operator $R(u, \alpha)$:

$$\tilde{f}_\alpha(t) = R(u_\delta, \alpha).$$

This operator is a regularizing operator for equation (6.0.1) and hence is stable.

We note that the value of the parameter α must be chosen so as to be compatible with the error in the initial data δ; that is, $\alpha = \alpha(\delta)$. It can be found, for example, from the condition $\rho_{l_2}(A\tilde{f}_\alpha, u_\delta) = \delta$, which can also be written in the form

$$\sum_{n=1}^{\infty} c_n^2 \frac{\alpha^2 \xi_n^2}{(1+\alpha\xi_n)^2} = \delta^2.$$

This is justified in exactly the same way as was described in Chapter II.

For $\delta^2 < \sum_{n=1}^{\infty} c_n^2$, the parameter α is uniquely determined since the function

$$\psi(\alpha) = \sum_{n=1}^{\infty} c_n^2 \frac{\alpha^2 \xi_n^2}{(1+\alpha\xi_n)^2}$$

is a continuous increasing function for $\alpha > 0$ and its value at $\alpha = 0$ is 0. To see the monotonicity, note that

$$\psi'(\alpha) = \sum_{n=1}^{\infty} c_n^2 \left\{ \frac{2\alpha\xi_n^2}{(1+\alpha\xi_n)^2} - \frac{\alpha^2\xi_n^2 \cdot 2\xi_n}{(1+\alpha\xi_n)^3} \right\} =$$

$$= 2\alpha \sum_{n=1}^{\infty} c_n^2 \frac{\xi_n^2}{(1+\alpha\xi_n)^3} > 0.$$

Since $\psi(\alpha) \leqslant \sum_{n=1}^{\infty} c_n^2$, the equation $\psi(\alpha) = \delta^2$ has no solution for $\delta^2 > \sum_{n=1}^{\infty} c_n^2$.

Remark 2. The sum of the original series

$$\sum_{n=}^{\infty} c_n \varphi_n(t),$$

understood as the limit

$$\lim_{k\to\infty} \sum_{n=1}^{k} c_n \varphi_n(t),$$

cannot serve as an approximation of the sum of the series

$$\sum_{n=1}^{\infty} c_n \varphi_n(t)$$

because of its instability under small changes (in the l_2-metric) in the coefficients c_n. On the other hand, the sum of the series

$$\sum_{n=1}^{\infty} \frac{1}{1+\alpha\xi_n} c_n \varphi_n(t),$$

understood again as the limit

$$\lim_{k\to\infty} \sum_{n=1}^{k} \frac{1}{1+\alpha\xi_n} c_n \varphi_n(t),$$

is stable under small changes (in the l_2-metric) in the coefficients c_n, and, for a value $\alpha(\delta)$ of α compatible with the error in the coefficients c_n, it approximates the function $\hat{\hat{f}}(t)$ uniformly.

Thus, the factors

$$r(n, \alpha) = \frac{1}{1+\alpha\xi_n}$$

play a stabilizing role. We shall call them **stabilizing factors.**

If we set

$$r(k, \alpha) = \begin{cases} 1 & \text{for } k \leqslant n, \\ 0 & \text{for } k > n, \end{cases}$$

we obtain the commonly used summation method mentioned on page 179. In this case, we need to take the sequence $\{\xi_k\}$ in the form

$$\xi_k < \infty \quad \text{for } k \leqslant n, \quad \xi_k = \infty \quad \text{for } k > n,$$

and then set $\alpha = 0$.

Remark 1. If in the classical method of summing a Fourier series by taking the limit as $n \to \infty$ of the sequence of partial sums

$$\sum_{k=1}^{n} c_k \varphi_k(t)$$

we take the number n compatible with the error δ in the sequence $\{c_n\}$, this way of summing will be stable.

Remark 2. If the function $\hat{f}(t)$ is piecewise-continuous, the method described is a stable summation method at every point of continuity of $\hat{f}(t)$.

§2. Optimal methods of summing Fourier series.

1. Since there is a large class of stable methods of summing Fourier series, it is natural to consider the problem of finding the one that is optimal in some respect or other. In the present section, we shall solve this problem in various formulations [12].

Approximately known coefficients in a Fourier series can contain uncontrollable random errors. Therefore, we can treat them as random numbers and use probabilistic methods in solving the problem of approximate summation of the series.

2. Let us assume that the errors in the Fourier coefficients, that is, the numbers γ_n, are random numbers satisfying the following two requirements:

1) $\{\gamma_n\}$ is a sequence of pairwise uncorrelated random numbers;

2) the mathematical expectations of these numbers are equal to 0; that is, $\bar{\gamma}_n = 0$ for all values of n.

Under these conditions, the approximate values of the Fourier coefficients c_n are also random numbers. The mathematical expectations of the random numbers c_n^2 and γ_n^2 are connected by

$$\overline{c_n^2} = a_n^2 + \overline{\gamma_n^2}.$$

The variances of the random variables c_n and γ_n are the same and are equal to $\sigma_n^2 = \overline{\gamma_n^2}$. We shall assume them to be known.

The functions $\tilde{f}_\alpha(t)$ minimizing the functional M^α for a fixed sequence $\{\xi_n\} \in \mathfrak{M}$ is therefore a random function.

Let us define

$$(\Delta f_\alpha)_\xi = \tilde{f}_\alpha(t) - \hat{f}(t),$$

where

$$\hat{f}(t) = \sum_{n=1}^{\infty} a_n \varphi_n(t).$$

For a measure of the deviation of $\tilde{f}_\alpha(t)$ from $f(t)$, we can take the mathematical expectation of the square of $(\Delta f_\alpha)_\xi$, that is, the quantity*

A) $$\overline{(\Delta f_\alpha)_\xi^2}$$

or the integral

B) $$\int_D \rho(t) \overline{(\Delta f_\alpha)_\xi^2}\, dt.$$

3. Thus, suppose that we have a Fourier series with approximate coefficients c_n. As we saw in §2, an approximate sum of it that is stable under small changes (in the l_2-metric) in the sequence $\{c_n\}$ depends on α and on the choice of sequence $\{\xi_n\}$ in $\mathfrak{M}$. Let us pose the problem of finding the approximate sum that deviates least (for fixed α) from $\hat{f}(t)$ in the sense of

$$\inf_{\{\xi_n\}\in\mathfrak{M}} \int_D \overline{(\Delta f_\alpha)_\xi^2} \rho(t)\, dt \quad \text{and} \quad \inf_{\{\xi_n\}\in\mathfrak{M}} \overline{(\Delta f_\alpha)_\xi^2}.$$

What supplementary information regarding the coefficients c_n and the noise must we have for this? To answer this question, we find the approximate sums that are optimal in the senses indicated. For a deviation of type B), we have

*If we know the probability distribution density $p(x)$ of the quantity $(\Delta f)_\xi^2 = x$, then the mathematical expectation $\overline{(\Delta f)_\xi^2})$ is calculated from the formula (for $a \leqslant x \leqslant b$)

$$\overline{(\Delta f)_\xi^2} = \int_a^b (\Delta f)_\xi^2 p(x)\, dx.$$

$$\int_D \overline{(\Delta f_\alpha)^2_\xi}\rho(t)\,dt = \sum_{n=1}^{\infty} \frac{\sigma_n^2 + \alpha^2\xi_n^2(\bar{c}_n^2 - \sigma_n^2)}{(1+\alpha\xi_n)^2} = \Phi(\xi_1, \ldots, \xi_n, \ldots).$$

The deviation will be minimum for those values of $\xi_n = \xi_n'$ at which the derivatives $\partial\Phi/\partial\xi_n$ vanish. From these conditions, we easily find that

$$\xi_n' = \frac{\sigma_n^2}{\alpha(\bar{c}_n^2 - \sigma_n^2)}.$$

Consequently,

$$r'(n, \alpha) = \frac{1}{1+\alpha\xi_n'} = 1 - \frac{\sigma_n^2}{\bar{c}_n^2}.$$

Thus, the sum that is optimal in the sense of the minimum of **B)** has the form

$$\tilde{f}_{\text{opt}}(t) = \sum_{n=1}^{\infty} \left(1 - \frac{\sigma_n^2}{\bar{c}_n^2}\right) c_n\varphi_n(t). \qquad (6.2.1)$$

Analogously, for a deviation of type A) we find that the sum that is optimal at a fixed point $t = t_0$ has the form

$$f_{\text{opt}}(t_0) = \sum_{n=1}^{\infty} \left(1 - \frac{\sigma_n^2}{\bar{c}_n^2}\right) c_n\varphi_n(t_0). \qquad (6.2.2)$$

Thus, to obtain sums of series with approximate coefficients c_n that are optimal in the senses indicated, we need to know the values of $\bar{c}_n^2$ and σ_n^2 for all values of n. This means that we need to know the exact Fourier coefficients a_n since* $\bar{c}_n^2 = a_n^2 + \sigma_n^2$.

*Compare with Wiener's optimal filtering in Chapter V, §2.

But only the ratios $\sigma_n^2/\bar{c}_n^2$ appear in formulas (6.2.1) and (6.2.2). In a number of specific problems, we can find approximate values of these ratios. How do the sums (6.2.1) and (6.2.2) then change?

Suppose that

$$\left(\frac{\sigma_n^2}{c_n^2}\right)_{\text{approx}} = \frac{\sigma_n^2}{c_n^2}(1+\beta_n),$$

where

$$\sum_{n=1}^{\infty} \beta_n^2 \leqslant \epsilon.$$

From these values, we obtain the sum

$$\tilde{f}_{\text{opt, approx}}(t) = \sum_{n=1}^{\infty}\left[1-\left(\frac{\sigma_n^2}{c_n^2}\right)_{\text{approx}}\right] c_n \varphi_n(t).$$

If we also know that, for all n,

$$\varphi_n^2(t) \leqslant Q,$$

then

$$|\tilde{f}_{\text{opt, approx}}(t) - \hat{f}(t)| \leqslant \sqrt{Q \cdot \epsilon \cdot \sum_{n=1}^{\infty} c_n^2}.$$

Thus, the optimal summation with stabilizing factors

$$r'(n, \alpha) = 1 - \sigma_n^2/\overline{c_n^2}$$

is stable under small changes in the numbers $\sigma_n^2/\overline{c_n^2}$ in the sense indicated.

Another method of stable summation of Fourier series is treated in [90].

CHAPTER VII

STABLE METHODS OF MINIMIZING FUNCTIONALS AND SOLVING OPTIMAL CONTROL PROBLEMS

1. A number of problems that are important in practice lead to mathematical problems of minimizing functionals $f[z]$. We need to distinguish between two kinds of such problems. The first kind includes problems in which we need to find the minimum (or maximum) value of a functional. Various problems of planning optimal systems or constructions are examples. With them, it is not important which elements z provide the sought minimum. Therefore, as approximate solutions we can take the values of the functional for any minimizing sequence $\{z_n\}$, that is, a sequence such that $f[z_n] \to \inf f[z]$ as $n \to \infty$.

The other kind includes problems in which we need to find the elements z that minimize the functional $f[z]$. We shall refer to these as problems of **minimization with respect to the argument**. Such problems include those in which the minimizing sequences may be divergent. In such cases, it is obvious that we cannot take as approximate solution the elements of the minimizing sequence. It is natural to call such problems unstable or ill-posed. Whole classes of optimal control problems that are important in practice are ill-posed in this sense. Examples are problems in which the functional to be optimized depends only on the phase variables [168].

In the present chapter, we shall look at stable methods of solving problems of the second kind.

2. Suppose that a continuous functional $f[z]$ is defined on a metric space F. The problem of minimizing $f[z]$ on F consists in finding an element $z_0 \in F$ that provides $f[z]$ with its smallest value f_0:

$$\inf_{z \in F} f[z] = f[z_0] = f_0. \tag{7.0.1}$$

Let us suppose that this problem has a unique solution z_0. Let $\{z_n\}$ denote a minimizing sequence, that is, one such that

$$\lim_{n \to \infty} f[z_n] = f_0.$$

We shall say that the minimization of the functional $f[z]$ on the set F is stable (see [171]) if every minimizing sequence $\{z_n\}$ converges (in the metric of the space F) to an element z_0 of F.

We shall say that the problem (7.0.1) of minimizing $f[z]$ on F is well-posed if it has a solution and is stable. Otherwise, we shall say that it is ill-posed.

Methods of direct minimization of the functional $f[z]$ are extensively used for finding the element z_0. With these methods, one constructs a minimizing sequence $\{z_n\}$ with the aid of some algorithm. Here, the elements z_n for which $f[z_n]$ are sufficiently close to f_0 are treated as approximate values of the element z_0 that one is seeking.

In their idea, direct minimization methods have great universality. Suppose, for example, that F is a set of functions $z(t)$ of a single variable t. Taking for the function $z(t)$ its grid approximation $z_i = z(t_i)$ for $i = 1, 2, \ldots, n$, we arrive at the problem of minimizing a function of n variables. The methods of solving such a problem are rather universal and not connected with any particular functional $f[z]$.

However, such an approach to the finding of an approximate solution is justifiable only when the minimizing sequence $\{z_n\}$ that is constructed converges to the element z_0.

We pointed out above that some functional-minimization problems do not have this property of stability. There exist minimizing sequences that do not converge to z_0.

To obtain approximate solutions of unstable problems of minimizing functionals $f[z]$, it is sufficient to give algorithms for constructing minimizing sequences $\{z_n\}$ that converge to an element z_0.

3. Let us consider solutions of differential equations of the form

$$dx/dt = F(t, x, u), \tag{7.0.2}$$

that satisfy the initial condition

$$x(t_0) = x_0, \tag{7.0.3}$$

where $x(t) = \{x^1(t), x^2(t), \ldots, x^n(t)\}$ is an n-dimensional-vector-valued function defined on an interval $t_0 \leqslant t \leqslant T$, x_0 is a given vector, and $u(t) = \{u^1(t), u^2(t), \ldots, u^m(t)\}$ is an m-dimensional-vector-valued function (the control) with range in an m-dimensional metric space U.

Let $f[x(t)]$ denote a given nonnegative functional (the target functional) defined on the set of solutions of the system (7.0.2).

Obviously, the solutions of the system (7.0.2) depend on the chosen control $u(t)$; that is, $x(t) = x_u(t)$. Therefore, the value of the functional $f[x(t)]$ for each solution of the system (7.0.2) is a functional of the controlling function $u(t)$. It is defined on the set U; that is, $f[x_u(t)] = \Phi[u]$.

The problem of optimal control can be formulated, for example, as the problem of finding, in some class U_1 of functions in the space U, a controlling function $u_0(t)$ that minimizes (or maximizes) the functional $\Phi[u] = f[x_u(t)]$.

Like the problem of minimizing the functional $f[z]$, which was described in subsection 1, this problem also can be unstable. Consider the case in which the class of admissible controls U_1 is the space of functions of a single variable t with the uniform metric and $f[x(t)]$ is a continuous functional. Then, for any

$\epsilon > 0$, there exists a control $u_1(t)$ such that

1) $$f[x_{u_1}(t)] \leqslant f[x_{u_0}(t)] + \epsilon,$$

and

2) the difference $u_1(t) - u_0(t)$ may assume arbitrarily large values allowed by membership of the functions $u_1(t)$ and $u_0(t)$ in the class U_1.

Let us take a function $u_1(t)$ that coincides with $u_0(t)$ everywhere except in the interval $(t_1 - \tau, t_1 + \tau)$ in which the difference $u_1(t) - u_0(t)$ exceeds a fixed number B allowed by membership of the functions $u_1(t)$ and $u_0(t)$ in the class U_1. Obviously, for any $\delta > 0$, there exists a $\tau = \tau(\delta)$ such that the inequality $|x_{u_1}(t) - x_{u_0}(t)| < \delta$ holds for solutions $x_{u_1}(t)$ and $x_{u_0}(t)$ of the system (7.0.2), (7.0.3) corresponding to the controls $u_1(t)$ and $u_0(t)$. By choosing δ and hence $\tau = \tau(\delta)$ sufficiently small, we can get $f[x_u(t)] \leqslant f[x_{u_0}(t)] + \epsilon$ whereas $\sup_t |u_1(t) - u_0(t)| \geqslant B$. This establishes the existence of unstable optimal control problems.

Remark. As target functional we can take a functional depending both on the phase variables and on the controlling functions.

In the present chapter, we shall look at algorithms for constructing minimizing sequences that converge to an element for which the functional in question attains its smallest value.

§ 1. A stable method of minimizing functionals.

1. Suppose that we are required to find an element z_0 of F at which a given functional $f[z]$ attains its smallest value on the set F:

$$\inf_{z \in F} f[z] = f[z_0] = f_0. \tag{7.1.1}$$

In this section, we shall, following [171], examine ways of

constructing minimizing sequences $\{z_n\}$ that converge to the element z_0.

We shall say that a minimizing sequence $\{z_n\}$ of the functional $f[z]$ is **regularized** if there exists a set $\tilde{F}$ which is compact in F and to which all the z_n belong [171].

Obviously, a regularized minimizing sequence converges to the element z_0 if the problem of minimizing the functional $f[z]$ has a unique solution z_0. If the functional $f[z]$ is bounded below, then the functional $f[z]$ attains its greatest lower bound at any limit point $\tilde{z}$ of the regularized minimizing sequence $\{z_n\}$; that is,

$$f[\tilde{z}] = f_0.$$

Having these minimizing sequence $\{z_n\}$, we can take, as approximate values of the element z_0 that we are seeking, elements z_n corresponding to those values of n for which $f[z_n]$ coincides with f_0 within the prescribed accuracy.

Thus, it is sufficient for us to give algorithms for constructing regularized minimizing sequences. This can be done by using the stabilizing functionals $\Omega[z]$ described in Chapter II. Let us give some definitions that we shall need [171].

2. Let $\tilde{F}$ denote a subset of F and suppose that $\Omega[z]$ is a stabilizing functional. Let us introduce a metric $\tilde{\rho}(z_1, z_2)$ for $z_1, z_2 \in \tilde{F} \subset F$ which majorizes the metric of F (that is, $\tilde{\rho}(z_1, z_2) \geqslant \rho_F(z_1, z_2)$ for all z_1 and z_2 in $\tilde{F}$) and for which the balls

$$S_r \equiv \{z;\, z \in \tilde{F},\, \tilde{\rho}(z, \tilde{z}_0) \leqslant r\}$$

in the space $\tilde{F}$ with arbitrary center $\tilde{z}_0$ are compact in F (in the metric of the space $\tilde{F}$). We shall say in this case that $\tilde{F}$ is **s-compactly embedded** in F.

If the metric $\tilde{\rho}(z_1, z_2)$ defines an s-compact embedding of $\tilde{F}$ in F and if $\varphi(q)$ is a nonnegative increasing function, then the functional

$$\Omega[z] = \varphi(\tilde{\rho}(z, \tilde{z}_0))$$

is obviously a stabilizing functional.

3. In calculating the functionals $f[z]$, one often uses approximate values of them. Approximate calculation of a functional $f[z]$ can be regarded as the calculation of another functional $\tilde{f}[z]$ the norm of whose deviation from $f[z]$ is small.

We shall refer to the smallest number δ for which the inequality

$$|\tilde{f}[z] - f[z]| \leqslant \delta\Omega[z]$$

holds for all elements z of $\tilde{F}$ as the **norm of the deviation**, with respect to the stabilizing functional $\Omega[z]$, between the functionals $\tilde{f}[z]$ and $f[z]$ defined on the set $\tilde{F}$.

4. Let us turn to the construction of regularized minimizing sequences of a functional $f[z]$. Let us suppose that

$$\inf_{z \in F} f[z] = f[z_0] = f_0.$$

Suppose that $\Omega[z]$ is a stabilizing functional and that $f_\delta[z]$ is a parametric family of functionals (defined for all $\delta \geqslant 0$) that approximate the functional $f[z]$ on the set $\tilde{F}$ in such a way that

$$|f_\delta[z] - f[z]| \leqslant \delta\Omega[z].$$

Let us assume that the element z_0 minimizing the functional $f[z]$ belongs to the set $\tilde{F}$.

For every $\alpha > 0$, consider the functional

$$M^\alpha[z, f_\delta] = f_\delta[z] + \alpha\Omega[z],$$

defined for every $z \in \tilde{F}$. Obviously, if $\delta < \alpha$, then $M^\alpha[z, f_\delta]$ is bounded below (on $\tilde{F}$) since

$$M^\alpha[z, f_\delta] \geqslant f[z] - \delta\Omega[z] + \alpha\Omega[z] \geqslant f_0.$$

Consequently, it has a greatest lower bound

$$M_{\alpha,\delta} = \inf_{z \in \tilde{F}} M^{\alpha}[z, f_{\delta}].$$

We shall say that an element $z_{\alpha,\delta}$ **almost minimizes** the functional $M^{\alpha}[z, f_{\delta}]$ if

$$M^{\alpha}[z_{\alpha,\delta}, f_{\delta}] \leqslant M_{\alpha,\delta} + \alpha\xi,$$

where $\xi = \xi(\alpha)$ and $0 \leqslant \xi(\alpha) \leqslant \xi_0 = \text{const}$.

Theorem 1. *For every $\epsilon > 0$, there exists an $\alpha_0 = \alpha_0(\epsilon)$ such that the inequality*

$$\tilde{\rho}_{\tilde{F}}(z_{\alpha,\delta}, z_0) \leqslant \epsilon$$

is satisfied for every element $z_{\alpha,\delta}$ that almost minimizes the functional $M^{\alpha}[z, f_{\delta}]$ with parameters $\alpha \leqslant \alpha_0(\epsilon)$ and $\delta/\alpha \leqslant q < 1$ if the metric in $\tilde{F}$ majorizes the metric of the space F.

Proof. To prove the theorem, it will be sufficient to show that, for any two sequences $\{\alpha_n\}$ and $\{\delta_n\}$ that converge to 0 and have the property that, for every n,

$$\delta_n/\alpha_n \leqslant q < 1,$$

the sequence $\{z_{\alpha_n,\delta_n}\}$ of elements z_{α_n,δ_n} that almost minimize the functionals $M^{\alpha_n}[z, f_{\delta_n}]$ converges to the element z_0. Obviously,

$$M^{\alpha_n}[z_{\alpha_n,\delta_n}, f_{\delta_n}] = f_{\delta_n}[z_{\alpha_n,\delta_n}] + \alpha_n\Omega[z_{\alpha_n,\delta_n}] \leqslant$$
$$\leqslant M_{\alpha_n,\delta_n} + \alpha_n\xi_0 \leqslant f_{\delta_n}[z_0] + \alpha_n\Omega[z_0] + \alpha_n\xi_0.$$

Using the inequalities

$$f_{\delta_n}[z] \geqslant f[z] - \delta_n\Omega[z],$$
$$f_{\delta_n}[z] \leqslant f[z] + \delta_n\Omega[z],$$

we obtain

$$f[z_{\alpha_n,\delta_n}] - \delta_n \Omega[z_{\alpha_n,\delta_n}] + \alpha_n \Omega[z_{\alpha_n,\delta_n}] \leqslant$$
$$\leqslant f[z_0] + (\alpha_n + \delta_n)\Omega[z_0] + \alpha_n \xi_0. \quad (7.1.2)$$

Since

$$f[z_{\alpha_n,\delta_n}] \geqslant f[z_0],$$

we obtain from (7.1.2)

$$(\alpha_n - \delta_n)\Omega[z_{\alpha_n,\delta_n}] \leqslant (\alpha_n + \delta_n)\Omega[z_0] + \alpha_n \xi_0.$$

Since $\delta_n \leqslant q\alpha_n$, if we now replace δ_n with $q\alpha_n$, we obtain

$$\Omega[z_{\alpha_n,\delta_n}] \leqslant \frac{1+q}{1-q}\Omega[z_0] + \frac{\xi_0}{1-q} = d_0.$$

Thus, all elements of the sequence $\{z_{\alpha_n,\delta_n}\}$ belong to the compact set

$$\widetilde{F}_{d_0} \equiv \{z; z \in \widetilde{F}, \Omega[z] \leqslant d_0\}.$$

Also, it follows from (7.1.2) that

$$0 \leqslant f[z_{\alpha_n,\delta_n}] - f[z_0] \leqslant (\alpha_n + \delta_n)\Omega[z_0] + \alpha_n \xi_0 \quad (7.1.3)$$

because

$$\delta_n - \alpha_n \leqslant \alpha_n q - \alpha_n = \alpha_n(q-1) < 0.$$

Since α_n and δ_n approach 0 as $n \to \infty$, it follows from (7.1.3) that the sequence $\{z_{\alpha_n, \delta_n}\}$ is a minimizing sequence for the functional $f[z]$. Since it belongs to the compact set $\widetilde{F}_{d_0}$, it is regularized and consequently it converges to z_0. This completes the proof of the theorem.

5. Let us look at the case in which $\widetilde{F}$ is a Hilbert space. We shall say that $\widetilde{F}$ is **s-compactly and continuously convexly embedded in F** if

a) the balls

$$S_r \equiv \{z;\, z \in \widetilde{F},\, \|z\| \leqslant r\}$$

are compact in F and

b) for any two sequences $\{z_n^{(1)}\}$ and $\{z_n^{(2)}\}$ of points in F such that $\rho_F(z_n^{(1)}, z_n^{(2)}) \to 0$ as $n \to \infty$, we have $\rho_F(z_n^{(1)}, \zeta_n) \to 0$ and $\rho_F(z_n^{(2)}, \zeta_n) \to 0$ as $n \to \infty$, where $\zeta_n = 0.5(z_n^{(1)} + z_n^{(2)})$.

If $\widetilde{F}$ is s-compactly embedded in F and $\varphi(q)$ is a nonnegative increasing continuous function, then, as was pointed out in subsection 2, a functional of the form

$$\Omega_1[z] = \varphi(\|z\|^2)$$

is a stabilizing functional.

Suppose that $f_\delta[z]$ is a δ-approximation of the functional $f[z]$ with respect to the stabilizing functional

$$\Omega_1[z] = \varphi(\|z\|^2).$$

For $\delta < \alpha$, the functional

$$M_1^\alpha[z, f_\delta] = f_\delta[z] + \alpha\Omega_1[z]$$

has a greatest lower bound $M_{\alpha,\delta}^1$ on the set $\widetilde{F}$:

$$M_{\alpha,\delta}^1 = \inf_{z \in \widetilde{F}} M_1^\alpha[z, f_\delta].$$

Theorem 2. *If $\widetilde{F}$ is a Hilbert space s-compactly and continuously convexly embedded in F, there exists an element $z_{\alpha,\delta}$ of $\widetilde{F}$ that minimizes the functional*

$$M_1^\alpha[z, f_\delta]$$

if $\delta/\alpha \leqslant q < 1$.

Proof. Suppose that $\{z_n\}$ is a sequence minimizing a functional $M_1^\alpha[z, f_\delta]$ such that the sequence $\{M_1^\alpha[z_n, f_\delta]\}$ converges to $M_{\alpha,\delta}^1$. Without loss of generality, we may assume that it is a decreasing sequence, so that, for every n,

$$M_1^\alpha[z_1, f_\delta] \geqslant f_\delta[z_n] + \alpha\Omega_1[z_n] \geqslant f_\delta[z_n] + (\alpha - \delta)\,\Omega_1[z_n].$$

Therefore,

$$\Omega_1[z_n] = \varphi(\|z_n\|^2) \leqslant \frac{1}{\alpha(1-q)}\{M_1^\alpha[z_1, f_\delta] - f[z_n]\}.$$

Since $f[z_n] \geqslant f[z_0]$, where z_0 is an element minimizing the functional $f[z]$, we have

$$\varphi(\|z_n\|^2) \leqslant \frac{1}{\alpha(1-q)}\{M_1^\alpha[z_1, f_\delta] - f[z_0]\} = C_1,$$

where the constant C_1 is independent of n. It follows that, for every n,

$$\|z_n\| \leqslant C_2,$$

where C_2 is a constant independent of n. Thus, $\{z_n\} \subset S_r$ for $r = C_2$. Since the balls S_r are assumed to be compact in F, the sequence $\{z_n\}$ is also compact in F. Without changing the notation, let us assume that this last sequence converges (in the metric of F) to an element $\bar{z}$ of F.

Let us show that it converges strongly to an element $\tilde{z}$ of $\widetilde{F}$. To do this, let us show that it is fundamental in $\widetilde{F}$, that is, that, for every $\epsilon > 0$, there exists an $n(\epsilon)$ such that the inequality

$$\|z_{n+p} - z_n\| \leqslant \epsilon$$

holds for $n \geqslant n(\epsilon)$ and $p > 0$. Let us suppose that this is not the case. Then, there exist an ϵ_0 and numerical sequences $\{n_k\}$ and $\{m_k\}$, where $m_k = n_k + p$, for which

$$\| z_{m_k} - z_{n_k} \| \geqslant \epsilon_0.$$

Let us define

$$\xi_k = 0.5\,(z_{m_k} - z_{n_k}).$$

Then,

$$\zeta_k = 0.5\,(z_{m_k} + z_{n_k}) = z_{n_k} + \xi_k = z_{m_k} - \xi_k.$$

Since z_{n_k} and z_{m_k} are elements of a minimizing sequence $\{z_n\}$ of the functional $M_1^\alpha[z, f_\delta]$ and since the sequence

$$\{M_1^\alpha[z_n, f_\delta]\}$$

is decreasing, it is obvious that, for sufficiently large k,

$$f_\delta[\zeta_k] + \alpha\varphi(\|\zeta_k\|^2) - \{f_\delta[z_{n_k}] + \alpha\varphi(\|z_{n_k}\|^2)\} \geqslant -\epsilon_k'$$

and

$$f_\delta[\zeta_k] + \alpha\varphi(\|\zeta_k\|^2) - \{f_\delta[z_{m_k}] + \alpha\varphi(\|z_{m_k}\|^2)\} \geqslant \epsilon_k'',$$

where ϵ_k and ϵ_k approach 0 as $k \to \infty$.

By virtue of the continuous convexity of the embedding of $\widetilde{F}$ in F,

$$\lim_{k\to\infty} f_\delta[z_{n_k}] = \lim_{k\to\infty} f_\delta[z_{m_k}] = \lim_{k\to\infty} f_\delta[\zeta_k]$$

and

$$\begin{aligned} &\varphi(\|\zeta_k\|^2) - \varphi(\|z_{n_k}\|^2) \geqslant -\bar{\epsilon}_k', \\ &\varphi(\|\zeta_k\|^2) - \varphi(\|z_{m_k}\|^2) \geqslant \bar{\epsilon}_k'', \end{aligned} \tag{7.1.4}$$

where $\bar{\epsilon}_k'$ and $\bar{\epsilon}_k''$ approach 0 as $k \to \infty$.

Since $\varphi(\|z\|^2)$ is an increasing function that is uniformly continuous in the region $\|z\| \leqslant C_2$, we obtain from (7.1.4)

$$\{\|z_{n_k}\|^2 + 2(z_{n_k}, \xi_k) + \|\xi_k\|^2\} - \|z_{n_k}\|^2 \geqslant -\beta_k'$$

and

$$\{\|z_{m_k}\|^2 - 2(z_{m_k}, \xi_k) + \|\xi_k\|^2\} - \|z_{m_k}\|^2 \geqslant -\beta_k'',$$

where β_k' and β_k'' approach 0 as $k \to \infty$. It follows that

$$2\|\xi_k\|^2 - 2(z_{m_k} - z_{n_k}, \xi_k) = -2\|\xi_k\|^2 \geqslant -(\beta_k' + \beta_k'')$$

or

$$\|\xi_k\|^2 \leqslant 0.5(\beta_k' + \beta_k'') \to 0 \quad \text{as} \quad k \to \infty,$$

which contradicts the assumption according to which

$$\|\xi_k\| = 0.5\|z_{m_k} - z_{n_k}\| \geqslant 0.5 \cdot \varepsilon_0.$$

Thus, $\{z_n\}$ is a fundamental sequence in $\widetilde{F}$ that converges strongly to an element $\tilde{z}$ of $\widetilde{F}$. Since the metric in $\widetilde{F}$ majorizes the metric in F, it follows that $\tilde{z} = \bar{z}$. Obviously, the element $z_{\alpha,\delta} = \bar{z} = \tilde{z}$ is the element that we are seeking, the one that minimizes the functional $F_1[z, f_\delta]$. This completes the proof of the theorem.

§2. A stable method of solving optimal-control problems.

1. Suppose that we are given a system of equations

$$dx/dt = F(t, x, u), \tag{7.2.1}$$

where $x(t) = \{x^1(t), x^2(t), \ldots, x^n(t)\}$ is an n-dimensional-vector-valued function defined on an interval $t_0 \leqslant t \leqslant T$ and $u(t) = \{u^1(t), u^2(t), \ldots, u^m(t)\}$ is a controlling vector-valued function with range in a complete m-dimensional metric space U. Let us look at solutions of the system (7.2.1) that satisfy the initial condition

$$x(t_0) = x_0, \tag{7.2.2}$$

where x_0 is a given vector.

Below, we shall look at the class of optimal-control problems in which the target functional depends only on the phase variables.

Let $f[x]$ denote a given nonnegative continuous functional defined on the solutions of the problem (7.2.1), (7.2.2). Since these solutions depend on u (so that we might write $x = x_u(t)$), the functional $f[x_u(t)] = \Phi[u]$ is defined on U. Let us look at the problem (problem U) of finding a control $u_0(t)$ minimizing the functional $\Phi[z]$. We shall call this control the optimal control. It was pointed out above that this problem is ill-posed. Let us suppose that there is an optimal control $u_0(t)$ in the space U.

A stable method for determining $u_0(t)$ approximately is described in the present section. Following [172], we shall do this by a regularization method. To obtain such a method, it is, as was mentioned above, sufficient to find an algorithm for constructing minimizing sequences $\{u_n(t)\}$ that converge to the function $u_0(t)$.

2. Consider the smoothing functional

$$B^\alpha[u] = \Phi[u] + \alpha\Omega[u],$$

where $\Omega[u]$ is a stabilizing functional. The functional $B^\alpha[u]$ is nonnegative. Therefore, it has a greatest lower bound B_0^α on U:

$$B_0^\alpha = \inf_{u \in U} B^\alpha[u].$$

Let U_1 denote a subset of U that admits a metrization $\rho_1(u_1, u_2)$ majorizing the metric of the space U.

Let us define $\Omega_1[u] = \rho_1^2(u, \bar{u}_0)$, $B_1^\alpha[u] = \Phi[u] + \alpha\Omega_1$, and $B_{01}^\alpha = \inf_{u \in U} B_1^\alpha[u]$, where $\bar{u}_0$ is a fixed element of U_1. Let $\{\alpha_n\}$

denote a decreasing sequence of positive numbers that converges to 0 and let $\{u_{\alpha_k}(t)\}$ denote a sequence of controls in U_1. Then we have

Theorem 1. *If problem U has a unique optimal control* $u_0(t)$ *belonging to the set* U_1, *then the sequence of functions* $\{u_{\alpha_k}(t)\}$ *satisfying the inequalities*

$$B^{\alpha_k}[u_{\alpha_k}(t)] \leqslant B_{01}^{\alpha_k} + \alpha_k C,$$

where C is a constant independent of α, *converges in the metric of U to* $u_0(t)$.

If U_1 is a Hilbert space, we have

Theorem 2. *If* U_1 *(with metric* $\rho_1(u_1, u_2)$*) is s-compactly and continuously convexly embedded in U, then there exists an element* $u_\alpha(t)$ *in* U_1 *that minimizes the functional* $B_1^\alpha[u]$.

The proofs of these theorems are completely analogous to the proofs of Theorems 1 and 2 of §1 and for this reason we omit them.

Obviously, these theorems lead directly to a stable method for determining the optimal control approximately.

Example. Let us look at the problem of the vertical ascent of a sounding rocket in a homogeneous atmosphere to its maximum altitude [31, 32]. For this we know the exact solution [127].

The vertical motion of a body of variable mass $m(t)$ in a homogeneous atmosphere is described by the system of equations

$$\frac{dv}{dt} = \frac{1}{m(t)}[au(t) - cv^2(t)] - g,$$

$$\frac{dm}{dt} = -u(t)$$

with initial conditions $m(0) = m_0$, $v(0) = v_0$. Here, $m(t)$ is the variable mass of the body ($m_0 \geqslant m(t) \geqslant \mu$, where μ is the mass of the rocket body), $v(t)$ is the velocity of the rocket, $u(t)$ is the controlling function (equal to the consumption of fuel per second during the time of flight), and a, c, and g are constants; specifically, a = 2500 m/sec is the velocity (relative to the rocket) at which the gases are ejected, $c = 0.2 \cdot 10^{-7}$ kg-sec/m is the generalized ballistic coefficient of air resistance and $g = 9.81\ \text{m/sec}^2$ is the acceleration due to gravity.

Even comparatively crude approximate solutions of this problem have a sharp burst at the initial instant ($t = 0$); that is, they have a δ-form nature. Therefore, it is natural to seek a solution in the form $u_1 = A\delta(t) + u(t)$, where A is a constant and $\delta(t)$ is Dirac's delta function. The function $u(t)$ that we are seeking is a continuous and sufficiently smooth function. It will be much simpler to find the function $u(t)$ numerically than to find a function with a δ-form singularity. Such a representation of the solution means that, to attain the maximum altitude, it is most convenient to burn up a certain amount of fuel instantaneously, and, after a certain velocity v_1^* is attained, to begin a gradual consumption of the fuel. Obviously, the optimal controlling function $\tilde{u}(t)$ must be positive on some interval $0 \leqslant t \leqslant T_1^*$ and equal to zero for $t > T_1^*$. We need to determine $\tilde{u}(t)$ and the parameters v_1^* and T_1^*.

Thus, we need to examine the two-parameter target functional $f[u, v_1, T_1]$ and, to do this, we need (using the condition that it be minimal) to find $\tilde{u}(t)$ and the parameters v_1^* and T_1^*.

According to Tsiolkovsky's formula, the mass m_1 of the fuel that must be consumed instantaneously to obtain a velocity v_1 is calculated from the equation

$$v_1 = v_0 + a\left|\ln\left(1 - \frac{m_1}{m_0}\right)\right|.$$

Then, A is determined from m_1. For what follows, let us take $v_0 = 0$ and $m_0 = 1$.

The altitude to which the rocket is lifted $H = H[v(u)]$ is equal to the integral $\int_0^T v(t)\,dt$, where T is the instant at which the velocity becomes equal to zero: $v(T) = 0$. It can be represented in the form $H = H_1 + \Delta H$, where

$$H_1 = \int_0^{T_1} v(t)\, dt, \quad \Delta H = \frac{\mu}{2c} \ln\left(1 + \frac{v_\mu^2 c}{\mu g}\right),$$

where in turn T_1 is the instant of termination of burning of fuel and v_μ is the velocity of the rocket at that instant.

As target functional, we take

$$f[u, v_1, T_1] = 1 - \frac{H[v(u)]}{H_0},$$

where H_0 is a number close to the maximum altitude that we are seeking. The problem of minimizing it is unstable. Let us solve it by the regularization method.

To find an approximate (regularized) solution, we choose a stabilizing functional $\Omega[u]$ of the form

$$\Omega[u] = \int_0^T (u'')^2\, dt.$$

The problem reduces to minimizing the functional

$$\Phi^\alpha[u, v_1, T_1] = f[u, v_1, T_1] + \alpha\Omega[u]$$

under the supplementary conditions $u(t) \geqslant 0$ and $\int_0^T u(t)\, dt = 1 - \mu$.

The procedure for finding an approximate solution of this problem for fixed α is as follows: we set up a sequence of pairs of numbers $\{v_1^{(n)}, T_1^{(n)}\}$. For each such pair we find a function $u_n^\alpha(t)$ minimizing the functional $\Phi^\alpha[u, v_1^{(n)}, T_1^{(n)}]$. Then, from the sequence $\{v_1^{(n)}, T_1^{(n)}\}$, we find a pair v_1^*, T_1^* that minimizes the functional

$$\Phi^{\alpha}\,[u_n^{\alpha}(t), v_1^{(n)}, T_1^{(n)}].$$

For large values of α (of the order of 10^5), the minimum of the functional $\Phi^{\alpha}[u, v_1, T_1]$ is attained with functions that are nearly constant. For small values of α (less than 0.1), the solution is very unstable under small random computational errors. For $\alpha = 10^3$, $v_1 = 9931.721$ m/sec, $T_1 = 176$ sec, and $\mu = 0.7$, we obtain the approximate optimal control shown in Figure 9 by the dashed curve. The solid line represents the exact optimal control. Here, H and H_1 have the values shown in the table.

The functional $\Phi^{\alpha}[u, v_1, T_1]$ was minimized by the method of projecting the gradient.

3. Let us look at a problem leading to minimization of functionals that is important in practice.

The inverse problem of antenna theory. Consider a linear antenna in the form of a straight rod of length $2l$. We direct the z-axis along the rod.

Suppose that a current of amplitude $j(z)$ varying with time according to the law $e^{i\omega t}$ is produced in the rod with the aid of special devices. This current causes an electromagnetic field symmetric about the z-axis in the space surrounding the rod. The electric and magnetic field intensities caused by the current $j(z)$

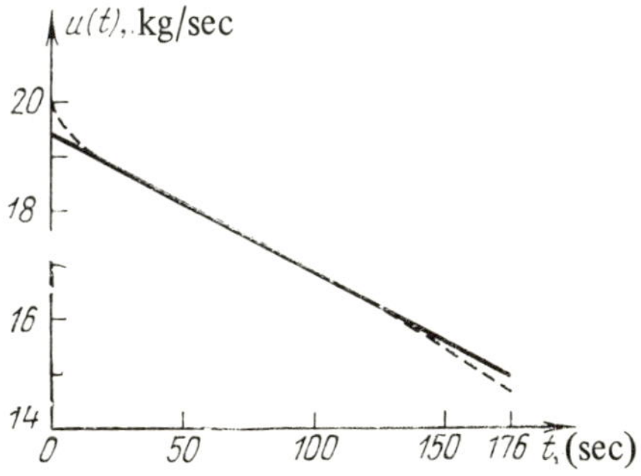

FIG. 9.

Exact solution	Approximation solution
$H_1=0.16507\cdot 10^7$ m $H=0.51496\cdot 10^7$ m	$H_1=0.16514\cdot 10^7$ m $H=0.51497\cdot 10^7$ m

and the intensity of the radiation of the antenna are determined by that current; that is, they are functionals $A_E[j]$, $A_H[j]$, and $A_M[j]$.

Let Q denote a plane in which the axis of the rod lies. Let us draw a circle in Q of radius R much greater than both the wave length λ and the rod length $2l$ with center at the midpoint of the rod. Then, at every point of that circle the magnetic field vector will be directed perpendicularly to the plane Q. Its magnitude will be a function of the radius R and the angle θ between the z-axis and the direction from the midpoint of the rod to the point of reception. Its relative value (relative, for example, to its value at some point on the circle) is a function only of the angle θ:

$$A[j]=P(\theta).$$

The function $P(\theta)$ is called the **directivity pattern** (with respect to the magnetic field intensity–the directivity pattern with respect to the electric field intensity or radiation flux is defined analogously).

In the case of a complicated configuration of the antenna, the directivity pattern $\vec{A}[j]$ will be in general be a vector-valued function of the two spherical coordinates θ and φ of the reception point:

$$A[\vec{j}]=\vec{P}(\theta,\varphi).$$

In the case of a linear antenna of length $2l$, the connection between the current $j(z)$ and the directivity pattern $P(\theta)$ is expressed by the formula (see [177])

$$A[j] \equiv \sin\theta \int_{-l}^{l} j(z)\, e^{-ikz\cos\theta} dz = P(\theta), \tag{7.2.3}$$

where $k = \omega/c$ is the wave number, c being the speed of light.

The direct problem in antenna theory consists in determining the directivity pattern $\vec{P}(\theta, \varphi)$ from a given current $\vec{j}(M)$, that is, in calculating the function $\vec{P}(\theta, \varphi) = A[\vec{j}]$.

What is usually meant by the inverse problem in antenna theory is the problem of finding the current distribution $\vec{j}(M)$ generating a given directivity pattern $\vec{P}(\theta, \varphi)$. For a linear antenna, this problem consists in solving a Fredholm integral equation of the first kind (7.2.3) for $j(z)$. As we know, this problem is ill-posed. It should be mentioned that equation (7.2.3) has a solution only for a very restricted class of right-hand members $P(\theta)$, that is, only for *realizable* directivity patterns. The given patterns usually do not belong to that class. Furthermore, a number of conditions (for example, boundedness of the current and of its derivative, minimum reactive power, etc.) are usually imposed in practice on the current distribution $\vec{j}(M)$, and it is not possible to allow for these in solving equation (7.2.3). Additional conditions are imposed on the directivity pattern. These include, for example, the requirement that it be close to a given directivity pattern, that it be nonzero in a fixed interval of angles but zero outside that interval (this corresponds to the case of strictly directed radio transmission), that it have its maximum value in the principal lobe, that the energy in the lateral lobes be minimum, etc. Therefore, it is expedient to formulate the inverse problem in a different manner.

Suppose that $\Phi_i[\vec{P}(\theta, \varphi)]$, for $i = 1, 2, \ldots, n$, are functionals such that the ith of them is defined on the set of functions $\vec{P}(\theta, \varphi)$ satisfying the ith condition imposed on the directivity patterns.

Suppose that $\Psi_k[\vec{j}(M)]$, for $k = 1, 2, \ldots, m$, are functionals defined on the set of functions $\vec{j}(M)$ satisfying the condition imposed on the current distribution. These functionals characterize the complexity of the construction.

Consider the functionals

$$\Psi[\vec{j}] = \sum_{k=1}^{m} \gamma_k \Psi_k [\vec{j}] \quad \text{and} \quad \Phi[\vec{P}] = \sum_{i=1}^{n} \beta_i \Phi_i [\vec{P}],$$

where the γ_k and the β_i are positive weighting coefficients chosen on the basis of the significance and influence of the corresponding conditions imposed on the current distribution and on the directivity patterns. The functional $\Phi[\vec{P}]$ is defined on the set of functions satisfying all the conditions imposed on the directivity patterns. The functional $\Psi[\vec{j}]$ is defined on the set of functions $\vec{j}(M)$ satisfying all the conditions imposed on the current distributions.

Therefore, following [177], what we shall consider as an approximate solution of the inverse problem of antenna theory is a current $\vec{j}(M)$ minimizing the functional Φ under the condition $\Psi \leqslant \Psi_0$ or the functional

$$\Phi[A[\vec{j}]] + \delta\Psi[\vec{j}],$$

where δ is a numerical factor characterizing the influence of the restrictions on the current distribution. Problems of minimizing such functionals have been examined above.

The definition of the optimal set of parameters is usually controlled by making a computational experiment.

This problem is a typical problem in the planning of optimal solutions (constructions).

CHAPTER VIII

STABLE METHODS OF SOLVING OPTIMAL-PLANNING (LINEAR-PROGRAMMING) PROBLEMS

Numerous optimal-planning and mathematical-programming problems (linear and nonlinear) are unstable: small changes in the initial data can lead to large (sometimes arbitrarily large) changes in the solution. In such problems, we are actually dealing with incomplete determinacy of the solution. This points to incompleteness in the formulation of the problem. Precision in formulating a problem is of great applied significance since the application of a mathematical tool to planning questions (to the development of optimal plans) is based on the solution of such problems.

Of special importance is the development of methods for finding "solutions" of such problems that are stable under small changes in the initial data.

The present chapter is devoted to these questions. In §1, we shall treat the formulation of optimal-planning and mathematical-programming problems (including linear-programming problems) and shall show how they are unstable. We shall introduce the concept of a normal solution. In the remaining sections, we shall look at the questions of existence and uniqueness of a solution and at methods, based on regularization (see Chapter II), for finding approximate solutions that are stable under small changes in the initial data.

§1. The formulation of optimal-planning and mathematical-programming problems.

1. Let us give the classical formulation of a typical optimal-planning problem. Let z_j, for $1 \leqslant j \leqslant n$, denote the number of objects of the jth kind that are produced, let α_j denote the maximum possible number of objects of the jth kind, and let c_j denote the total labor expended on the jth object at all the basic divisions of the plant. Then, the scalar product $\varphi_1(z) = (c, z)$ of the vectors $c = \{c_1, c_2 \ldots, c_n\}$ and $z = \{z_1, z_2, \ldots, z_n\}$ characterizes the load of the plant in the plan for producing z objects. Let b_i, for $1 \leqslant i \leqslant m$, denote the reserve of time of the ith division of the plant. Let a_{ij} denote the labor expended on the jth object in the ith division.

We denote by G the set of vectors z (plans) satisfying the conditions $b - Az \geqslant 0$ and $0 \leqslant z \leqslant \alpha$, where α and b are the vectors $\alpha = \{\alpha_1, \alpha_2, \ldots, \alpha_n\}$ and $b = \{b_1, b_2, \ldots, b_m\}$ and $A = \{a_{ij}\}$ is the matrix with elements a_{ij}.

The problem of determining the optimal plan $\bar{z}$ can consist in finding a vector $\bar{z}$ out of the set of vectors (plans) G for which the load of the plant is maximum, that is,

$$\max_{z \in G} \psi_1(z) = \psi_1(\bar{z}).$$

The function $\varphi(z)$ is called the **target function** of the problem.

The well-known simplex method was used to calculate from the data of one of the factories the optimal quarterly plans corresponding to some set of initial data with varying accuracy [178]. The results of the calculations are shown in Table 1, where the Roman numbers represent the variants of the solutions (corresponding to different initial data) and the Arabic numbers represent the components of the solutions (the vectors $\bar{z}$). It is obvious from the table that, for comparatively close optimal values of the target function $\varphi(\bar{z})$ (for differences of the order of 1%), the number of objects to be produced according to these optimal plans fluctuates from one kind of object to the other over a range of several hundreds. Thus, this problem is unstable.

TABLE 1

$\bar{z}$	I	II	III	IV	V	VI	VII	VIII	IX	X	XI
1	614	590	596	765	638	507	469	446	373	642	383
2	638	634	634	684	644	611	604	548	583	444	581
3	418	424	423	376	412	446	555	479	479	479	479
4						49	49				
5	66			36		105		66		232	238
6		36					60		106		
7								105			
10			56								
11											
12	296	297	294	314	298	293	295	291	255	286	237
13	50	49	50		47	61	59	69	68	68	78
14									54		72
$\varphi(z)$ (1000 ruble)	2391	2390	2390	2388	2387	2382	2380	2376	2373	2371	2370

2. Optimal-planning problems constitute a special case of linear-programming problems, which in turn constitute a special case of mathematical-programming problems.

Let R^n denote the n-dimensional space of vectors $z = (z_1, z_2, \ldots, z_n)$ and let $g(z)$ and $h(z)$ denote given vector-valued functions defined on R^n:

$$g(z) = \{g_1(z), g_2(z), \ldots, g_p(z)\},$$
$$h(z) = \{h_{p+1}(z), h_{p+2}(z), \ldots, h_m(z)\} \quad (p < m),$$

where $g_i(z)$ and $h_j(z)$ are scalar functions.

We denote by G the set of vectors z in the space R^n for which $g(z) \geqslant 0$ and $h(z) = 0$:

$$G = \{z;\ g(z) \geqslant 0,\ h(z) = 0\}.$$

Let $\varphi(z)$ denote a given scalar function.

The mathematical-programming problem consists in finding a vector $\tilde{z}$ in R^n that minimizes the function $\varphi(z)$ on the set G:

$$\varphi(\tilde{z}) = \min_{z \in G} \varphi(z). \tag{8.1.1}$$

The function $\varphi(z)$ is called the **target function** of the problem (8.1.1).

If in this problem the functions $\varphi(z)$, $g(z)$, and $h(z)$ are linear, the problem is called a **linear-programming problem**. As a rule, optimal-planning problems fall in this class.

3. In practice, information regarding the functions $\varphi(z)$, $g(z)$, and $h(z)$ is approximate. For example, these functions are known with an error; hence, instead of them we can take any functions $\varphi_\delta(z)$, $g_\delta(z)$, and $h_\delta(z)$ for which*

---

*For simplicity of notation, we took δ to be the same for all three functions. In practice, an estimate of the error is determined by vectors $\delta' = (\delta_1', \delta_2', \ldots, \delta_p', \delta_{p+1}', \ldots, \delta_m')$ and $\delta'' = (\delta_1'', \delta_2'', \ldots, \delta_n'')$ on the basis of an appropriate choice of norms for them.

$$\|\varphi_\delta(z) - \varphi(z)\| \leqslant \delta, \quad \|g(z) - g_\delta(z)\| \leqslant \delta,$$
$$\|h_\delta(z) - h(z)\| \leqslant \delta.$$

The choice of the functions $\varphi_\delta(z)$, $g_\delta(z)$, and $h_\delta(z)$ from the set

$$Q_\delta(\varphi, g, h) \equiv \{(\varphi_\delta, g_\delta, h_\delta) : \|\varphi_\delta - \varphi\| \leqslant \delta,$$
$$\|g_\delta - g\| \leqslant \delta, \ \|h_\delta - h\| \leqslant \delta\}$$

is usually of a random nature.

Thus, we can make a judgment regarding the solution of the original problem (8.1.1) only from the solution of the approximate problem

$$\min_{z \in G_\delta} \varphi_\delta(z), \tag{8.1.2}$$

where $G_\delta = \{z;\ g_\delta(z) \geqslant 0,\ h_\delta(z) = 0\}$, which is randomly chosen from the class of approximate problems defined by Q_δ.

4. Let $H_\delta(\varphi, g, h)$ denote the set of solutions z_δ of problems of the form (8.1.2) for all $(\varphi_\delta, g_\delta, h_\delta)$ in Q_δ. The problem (8.1.1) is said to be **stable** if

$$\Delta_\delta = \sup_{z'_\delta, z''_\delta \in H_\delta} \|z'_\delta - z''_\delta\| \to 0 \quad \text{as} \quad \delta \to 0.$$

On the other hand, if there exists a number $B > 0$ such that, for every $\delta > 0$, there exists two triples $(\varphi'_\delta, g'_\delta, h'_\delta)$ and $(\varphi''_\delta, g''_\delta, h''_\delta)$ in Q_δ and solutions z'_δ and z''_δ of the problem (8.1.2) corresponding to them such that

$$\|z'_\delta - z''_\delta\| \geqslant B,$$

then the problem (8.1.1) is said to be **unstable**.

It is also natural to call unstable optimal-planning and mathematical-programming problems **ill-posed**. Obviously, the exact solutions z'_δ and z''_δ of the ill-posed problem (8.1.1) with approximate initial data $(\varphi'_\delta, h'_\delta, g'_\delta)$ and $(\varphi''_\delta, h''_\delta, g''_\delta)$ do not provide sufficient information regarding the solution of the original problem.

Thus, finding the exact solutions of problems of this kind cannot serve as a reliable method of solving optimal-planning problems with approximate initial data.

Such a situation often arises in computational work when one is using several methods of solving mathematical-programming problems. We mention that, despite the great differences in the solutions z'_δ and z''_δ, the values of the target functions $\varphi'_\delta(z'_\delta)$ and $\varphi''_\delta(z''_\delta)$ may differ only slightly from each other. What has been said is reflected in the results shown in Table 1.

§2. Optimal-planning problems. Existence and uniqueness of solutions.

1. Suppose that we are given a matrix $A = \{a_{ij}\}$ with elements a_{ij} and vectors $\bar{u} = \{\bar{u}_i\}$, for $i = 1, 2, \ldots, m$, and $c = \{c_j\}$, where $c_j \geqslant 0$, for $j = 1, 2, \ldots, n$. Let us look at the following optimal-planning problem.

In the set G of elements (vectors) $z = \{z_j\}$ in n-dimensional space R^n that satisfy the conditions

$$Az = \bar{u}, \tag{8.2.1}$$

$$z_j \geqslant 0 \quad (j = 1, 2, \ldots, n), \tag{8.2.2}$$

find an element (vector) $\bar{z} = \{\bar{z}_j\}$ minimizing the function

$$\varphi(z) = (c, z) = \sum_{j=1}^{n} c_j z_j, \tag{8.2.3}$$

that is, an element $\bar{z}$ such that

$$\varphi(\bar{z}) = \min_{z \in G} \varphi(z). \tag{8.2.4}$$

Obviously, this is a linear-programming problem. Here $g(z) = z$ and $h(z) = Az - \bar{u}$.

2. If the condition (system of equations) $Az = \bar{u}$ has linearly dependent rows, the problem (8.2.1)–(8.2.3) is in general ill-posed. One usually assumes that the rows of the given conditions are linearly independent. However, this assumption cannot be checked in practice when the initial data are given approximately. Therefore, we shall not assume linear independence of the given conditions and shall deal with ill-posed problems of the form (8.2.4).

Let us prove the existence of a solution to such problems.

Theorem 1. If conditions (8.2.1) and (8.2.2) are consistent, the problem (8.2.1)–(8.2.3) has at least one solution.

Proof. Suppose that $c_j > 0$ for $1 \leqslant j \leqslant j_0$ and $c_j = 0$ for $j_0 < j \leqslant n$. We denote by R_1 the set of vectors in the space R^n for which condition (8.2.2) holds:

$$R_1 = \{z;\ z_j \geqslant 0 \ \text{for} \ j = 1, 2, \ldots, n\}.$$

Let R_2 denote the set of vectors in R^n satisfying conditions (8.2.1) and (8.2.2):

$$R_2 \equiv \{z;\ z \in R_1,\ Az = u\}.$$

The assumed consistency of conditions (8.2.1) and (8.2.2) implies that R_2 is nonempty.

Let φ_0 denote the greatest lower bound of the set of values assumed on the set R_2 by the target function $\varphi(z)$ defined by (8.2.3) and let $\{z^{(k)}\}$ denote a minimizing sequence of points in R_2, that is, a sequence such that $\varphi(z^{(k)}) \to \varphi_0$ as $k \to \infty$. Obviously, we can assume that $\varphi(z^{(k)}) \leqslant \varphi(z^{(k-1)})$ for $k > 1$.

Since

$$\varphi(z^{(k)}) \leqslant \varphi(z^{(k-1)}),$$

it follows that, for $1 \leqslant j \leqslant j_0$,

$$c_j z_j^{(k)} \leqslant \varphi(z^{(k)}) \leqslant \varphi(z^{(1)}).$$

Consequently, for $1 \leqslant j \leqslant j_0$, the coordinates $z_j^{(k)}$ of the vector $z^{(k)}$ are bounded:

$$z_j^{(k)} \leqslant \frac{\varphi(z^{(1)})}{c_j}.$$

For all positive numbers d_j, the set of points in the space R^{j_0} for which

$$0 \leqslant z_j \leqslant d_j, \quad 1 \leqslant j \leqslant j_0,$$

is compact in R^{j_0}. The sequence $\{\tilde{z}^{(k)}\}$ of vectors

$$\tilde{z}^{(k)} = \{z_j^{(k)}\}, \quad 1 \leqslant j \leqslant j_0, \ k = 1, 2, \ldots$$

in the space R^{j_0} is contained in the parallelepiped with edges

$$d_j = \frac{\varphi(z^{(1)})}{c_j}.$$

Consequently, it has a subsequence that converges to a vector $\tilde{z} \in R^{j_0}$. Without changing the notation, let us suppose that

$$\lim_{k \to \infty} \tilde{z}^{(k)} = \tilde{z} = \{\tilde{z}_j\}, \ 1 \leqslant j \leqslant j_0.$$

Since $\varphi(z)$ depends only on the first j_0 coordinates of the vector z ($c_j = 0$ for $j > j_0$), it follows that $\varphi(\tilde{z}) = \varphi_0$.

Consider the system of equations in z''

$$A''z'' = -A'z' + \bar{u}, \tag{8.2.5}$$

where

$$z' = \{z_j\},\ 1 \leqslant j \leqslant j_0,\ z'' = \{z_j\},\ j_0 < j \leqslant n,$$

$$A' = \{a_{ij}\},\ 1 \leqslant j \leqslant j_0,\ A'' = \{a_{ij}\},\ j_0 < j \leqslant n,\ i = 1, 2, \ldots, m.$$

Since the system

$$A''z'' = \bar{u} - A'\tilde{z}^{(k)}$$

is consistent (suppose that $z'' = z_2^{(k)}$ is its solution) and since the vector $u = \bar{u} - A'\tilde{z}^{(k)}$ satisfies all the linear relationships that the rows of the matrix A'' satisfy, the system (8.2.5) is also consistent. It has a nonzero solution for which $z_j \geqslant 0$ (where $j_0 < j \leqslant n$).

To see this, suppose that it is not the case. Then, a linear manifold of elements z'' for which $A''z'' = u$ is at a finite positive distance from the set $R_1'' = \{z''; z_j \geqslant 0, j_0 < j \leqslant n\}$. In this case, the manifolds defined by

$$A''z'' = \bar{u} - A'\tilde{z}^{(k)} \qquad (k = 1, 2, \ldots)$$

for sufficiently large k are also at finite distances from R_1''. But this contradicts the membership of $\tilde{z}^{(k)}$ in R_1.

Taking any nonzero solution $z'' \in R_1''$ of the system (8.2.5) such that $z_j'' \geqslant 0$ (for $j_0 < j \leqslant n$), we see that the vector

$$\tilde{\tilde{z}} = (\tilde{z}_1, \tilde{z}_2, \ldots, \tilde{z}_{j_0}, z''_{j_0+1}, \ldots, z''_n)$$

satisfies conditions (8.2.1) and (8.2.2) and that $\varphi(\tilde{\tilde{z}}) = \varphi_0$. This completes the proof of the theorem.

The following example shows that the solution of the problem (8.2.1)–(8.2.3) may fail to be unique.

Example 1. Suppose that $\varphi(z) = z_3$ and that condition (8.2.1) has the form

$$z_1 - z_2 = 0.$$

Obviously, the minimum of the function $\varphi(z) = z_3$ on the set $R_1 = \{z; z_j \geqslant 0; j = 1, 2, 3\}$ is equal to 0 and it is attained at points of the ray $z_1 \geqslant 0$ defined by

$$z_2 = z_1, \qquad z_3 = 0.$$

3. When we have a set of solutions, to make the problem definite we need to impose supplementary conditions on the solution sought.

Suppose that it is a question of optimal-planning problems. Let us suppose that work is being performed in accordance with a plan $z^{(0)}$ and that it is necessary to modify this plan with change in the initial data. Different optimal plans correspond to the new initial data. It is natural to choose whichever plan deviates the least from the original plan $z^{(0)}$. A similar choice criterion is connected with the minimum expenditure on organizational readjustments that were not taken into account in the formulation of the problem. We take for measure of deviation between the new and old plans $\bar{z}^{(0)}$ and $z^{(0)}$ the weighted square deviation

$$\| \bar{z}^{(0)} - z^{(0)} \| = \left\{ \sum_j p_j \left(\bar{z}_j^{(0)} - z_j^{(0)}\right)^2 \right\}^{1/2}.$$

or, in general, some positive-definite quadratic form. With this in mind, we give the

Definition. Let $z^{(0)}$ denote a vector in R^n. We shall say that $\bar{z}^{(0)}$ is a **normal solution** of the problem (8.2.1)–(8.2.3) (with respect to $z^{(0)}$) if

$$\| \bar{z}^{(0)} - z^{(0)} \| \leqslant \| z - z^{(0)} \|,$$

where z is any solution of that problem. If the solution of the problem (8.2.1)–(8.2.3) is unique, it obviously coincides with the normal solution. If the problem has more than one solution, the existence of a normal solution is obvious since the set H of vectors minimizing the function $\varphi(z)$ is the intersection of the three closed sets

$$\{z;\ Az=\bar{u}\},\qquad R_1\equiv\{z;\ z_j\geqslant 0,\ j=1,2,\ \ldots,\ n\},$$
$$\{z;\ \varphi(z)=\varphi_0\}$$

and is hence itself closed.

Theorem 2. *The normal solution of the problem* (8.2.1)–(8.2.3) *is unique.*

Proof. Let us suppose that there exist two distinct normal solutions $z^{(1)}$ and $z^{(2)}$. Then, for any positive number α, the vector

$$\hat{z}=\alpha z^{(1)}+\beta z^{(2)},\qquad \beta=1-\alpha$$

satisfies condition (8.2.1) (by virtue of its linearity) and (8.2.2). Furthermore,

$$\varphi(\hat{z})=\alpha\varphi(z^{(1)})+\beta\varphi(z^{(2)})=\alpha\varphi_0+\beta\varphi_0=\varphi_0.$$

Since $z^{(1)}$ and $z^{(2)}$ are normal solutions, we have

$$\|z^{(1)}-z^{(0)}\|=\|z^{(2)}-z^{(0)}\|.$$

On the other hand, for $\alpha = 0.5$ we have

$$\hat{z}=0.5\,(z^{(1)}+z^{(2)})$$

and

$$\|\hat{z}-z^{(0)}\|^2\quad\|z^{(1)}-z^{(0)}\|^2-\|0.5\,(z^{(1)}-z^{(2)})\|^2.$$

Thus,

$$\|\hat{z}-z^{(0)}\|^2<\|z^{(1)}-z^{(0)}\|^2,$$

which contradicts the fact that $z^{(1)}$ is a normal solution if $z^{(1)}\neq z^{(2)}$. This completes the proof of the theorem.

It was mentioned above that the solution of the problem (8.2.1)–(8.2.3) is in general unstable under small changes in the original information. The following section is devoted to a stable method of solving such problems, namely, the regularization method.

§3. The regularization method of solving optimal-planning problems.

1. Usually, the initial data in optimal-planning problems are only approximate. In what follows, we shall distinguish between the solvable problem (the exact problem) and the problem actually assigned (given in a form approximating the exact one).

The assigned problem does not enable us to draw any conclusion either as to the stability of the solvable problem or as to the uniqueness of its solution even though that problem has these properties. As we saw, the exact solution of the given problem is ineffective for investigating the problem to be solved. From the point of view of the information available to us, any set of initial data $\{\varphi, h, g\}$ satisfying the conditions

$$\|\varphi_\delta(z) - \varphi(z)\| \leqslant \delta, \quad \|h_\delta(z) - h(z)\| \leqslant \delta,$$
$$\|g_\delta(z) - g(z)\| \leqslant \delta. \qquad (8.3.1)$$

can serve as initial data of the (exact) problem to be solved.

It should be noted that if in addition to the conditions $Az = u$ we add on some linearly dependent equations, this makes the problem unstable (even if it was stable beforehand) although it remains equivalent to that problem in the practical sense. For linear-programming (optimal-planning) problems with a sufficiently great number of conditions $Az = u$, we cannot as a rule actually test whether the condition of linear independence is satisfied or not.

Thus, an approach to the solution of linear-programming (optimal-planning) problems that does not require an assumption of linear independence of the conditions $Az = u$ is necessary. The present section is devoted to an exposition of a stable method of approximate determination of a normal solution to the exact problem.

2. The optimal-planning problem thus becomes a variational problem: Find the minimum of the function $\varphi(z)$ for $\varphi(z) \geqslant 0$) or, what amounts to the same thing, of the function $\varphi^2(z)$ on the set

$$G = \{z;\ z \in R_1,\ Az = u\}.$$

If $\varphi^2(z)$ is a stabilizing functional for that set, that is, if the set G_d of elements z of G for which $\varphi^2(z) \leqslant$ d is compact, then the existence of an element z_0 minimizing $\varphi(z)$ is obvious. However, as the example given above shows, $\varphi^2(z)$ is not always a stabilizing functional.

Let us look in greater detail at the definition of the measure of error in prescribing $\varphi(z)$. Suppose that a stabilizing functional $\Omega[z]$ (for example, $\Omega[z]$ might be $\sum_{i=1}^{n} p_i (z_i - z_i^0)^2)$ and target functions $\varphi(z)$ and $\tilde{\varphi}(z)$ are defined on R^n. We define the measure of deviation between $\varphi^2(z)$ and $\tilde{\varphi}^2(z)$ as the smallest number δ such that

$$|\varphi^2(z) - \tilde{\varphi}^2(z)| \leqslant \delta (1 + \Omega[z]).$$

For linear-programming problems, where $\varphi^2(z)$ and $\tilde{\varphi}^2(z)$ are quadratic functionals, such a definition is natural. The presence of the 1 in the right-hand member is necessary since $\Omega[z_0]$ can be equal to zero without having $\varphi^2(z_0)$ equal to zero.

3. Suppose that, in an optimal-planning problem, we are given not the exact initial data $\{A, \bar{u}, \bar{c}\}$ but δ-approximations of them $\{A, \tilde{u}, \tilde{c}\}$ such that

$$\|\tilde{u} - \bar{u}\| \leqslant \delta \quad \text{and} \quad |\varphi^2(z) - \tilde{\varphi}^2(z)| \leqslant \delta (1 + \Omega[z]),$$

where $\varphi(z) = (c, z)$, $\tilde{\varphi}(z) = (\tilde{c}, z)$, and $\Omega[z]$ is a stabilizing functional on R^n (a positive-definite quadratic form).

Let us take the auxiliary target function

$$\Phi^2(z) = \tilde{\varphi}(z) + \lambda(1 + \Omega[z]), \ (\lambda > 0)$$

and let us solve approximately the optimal planning problem with target function $\Phi^2(z)$. This substitution is admissible from the point of view of accuracy of giving of the target function if $\lambda \leqslant \delta$, since

$$|\Phi^2(z) - \tilde{\varphi}^2(z)| = \lambda(1 + \Omega[z]) \leqslant \delta(1 + \Omega[z]).$$

Thus, among the vectors z belonging to the set R_1 such that $\|Az - \tilde{u}\| \leqslant \delta$, we are required to find the vector z_δ minimizing the function $\Phi^2(z)$. Obviously, the auxiliary target function $\Phi^2(z)$ is a quasimonotonic stabilizing functional on the set R_1 (where $\Phi^2(z)$ is a quadratic functional).

Consequently, the conditions of the lemma of §2 of Chapter II are satisfied. According to that lemma, the greatest lower bound of the functional (function) $\Phi^2(z)$ is attained with that vector z_δ in R_1 for which $\|Az_\delta - \tilde{u}\| = \delta$. This problem is equivalent to the problem of minimizing the quadratic form

$$M_\lambda^\alpha[z, \tilde{u}, \tilde{c}, A] = \|Az - \tilde{u}\|^2 + \alpha[\tilde{\varphi}^2(z) + \lambda(1 + \Omega[z])]$$

with determination of the parameter α from the condition $\|Az_\alpha - \tilde{u}\| = \delta$, where z_α is a vector minimizing $M_\lambda^\alpha[z, \tilde{u}, \tilde{c}, A]$. Since $\|Az_\alpha - \tilde{u}\|^2$ is a quadratic form, the discrepancy $\varphi(\alpha) = \|Az_\alpha - \tilde{u}\|^2$ is a continuous increasing function, so that the parameter α is uniquely determined from the condition $\varphi(\alpha) = \delta^2$.

4. In the mathematical formulation of optimal-planning problems, one usually does not take into account all the factors in choosing the target function $\varphi(z) = (c, z)$. Such factors may include, for example, the requirement of only small deviation of the optimal plan sought, corresponding to new (and only slightly different from the original) initial data, from the preceding plan (the requirement of minimum organizational readjustment). The introduction of the term $\lambda(1 + \Omega[z])$ into the new target function

$\Phi^2(z)$ can be regarded as a correction for the influence of factors not taken into account in the target function $\tilde{\varphi}^2(z)$, and λ can be regarded as the value of an expert estimate of their influence.

Analogous motivation (see Chapter II, §2, subsection 8, where the problem $Az = u$ with approximately known operator $\tilde{A}$ and right-hand member $\tilde{u}$ is examined) leads to the construction of an approximate solution of the problem of optimal planning with approximate initial data $\{\tilde{A}, \tilde{u}, \tilde{c}\}$ as the solution of the problem of minimizing the smoothing functional

$$M_\lambda^\alpha[z, \tilde{u}, \tilde{c}, \tilde{A}] = \| \rho_1^2 + \alpha \{\tilde{\varphi}^2(z) + \lambda (1 + \Omega[z])\},$$

where α is defined on the basis of the discrepancy by

$$\rho_1^2 \equiv \rho_U^2 (\tilde{A}z_\alpha, \tilde{u}) - h^2 \Phi^2 (z) = \delta^2.$$

Here, h characterizes the error in assigning the operator $\tilde{A}$ (see §2 of Chapter II).

5. Let us look again at the linear-programming problem of finding an element $\bar{z}^0$ minimizing the function $\varphi(z) = (c, z)$ on the set

$$R_2 = \{z;\ Az = \bar{u},\ z \in R_1\},$$

where A is a linear operator.

Let z_0 denote an element relative to which we seek a normal solution. Consider the auxiliary problem I_λ: Find an element z_λ minimizing the functional

$$\Phi^2(z) = \varphi^2(z) + \lambda\Omega[z]$$

on the set R_2. Here, $\Omega[z]$ is a positive-definite quadratic form.

The existence of an element z_λ is obvious if the conditions defining R_2 are compatible. One can easily show that the element z_λ is unique. Specifically, for linear-programming problems, the set R_2 is convex. Suppose that there exist two elements z_λ^1 and z_λ^2 minimizing the quadratic functional $\Phi^2(z)$. On that segment of the line

$$z = z_\lambda^1 + \beta (z_\lambda^2 - z_\lambda^1), \quad -\infty < \beta < \infty,$$

belonging to R_2, the values of the function $\Phi^2(z)$ represent a quadratic function of β and thus cannot have two minimum values.

Above, we examined the optimal-planning problem in which the functional $\tilde{\varphi}^2(z)$ was replaced with

$$\Phi^2(z) = \tilde{\varphi}^2(z) + \lambda(1 + \Omega[z]).$$

Let us show that the solution z_λ of the problem with functional $\Phi^2(z)$ approaches the normal solution of the original problem as $\lambda \to 0$.

Let $\bar{z}^{(0)}$ denote the normal solution of the problem (8.2.1)–(8.2.3).

Theorem 1. *For every* $\epsilon > 0$, *there exists a* $\lambda_0(\epsilon)$ *such that*

$$\| z_\lambda - \bar{z}^{(0)} \| \leqslant \epsilon$$

for $\lambda \leqslant \lambda_0(\epsilon)$; *that is,* z_λ *approaches the normal solution of the problem (8.2.1)–(8.2.3) as* $\lambda \to 0$.

Proof. Let us suppose that this is not the case. Then, there exists an $\epsilon_0 > 0$ and a sequence $\{\lambda_k\}$ converging to zero such that $\| z_{\lambda_k} - \bar{z}^{(0)} \| \geqslant \epsilon_0$ for all k. Since z_{λ_k} minimizes the functional $\varphi^2(z) + \lambda_k \Omega[z]$, we have

$$\varphi^2(z_{\lambda_k}) + \lambda_k \Omega[z_{\lambda_k}] \leqslant \varphi^2(\bar{z}^{(0)}) + \lambda_k \Omega[\bar{z}^{(0)}].$$

Therefore,

$$\Omega[z_{\lambda_k}] \leqslant \Omega[\bar{z}^{(0)}] + \frac{\varphi^2(\bar{z}^{(0)}) - \varphi^2(z_{\lambda_k})}{\lambda_k}.$$

Since $\varphi(\bar{z}^{(0)}) \leqslant \varphi(z)$ for all z in the set $R_2 = \{z; Az = \bar{u}, z \in R_1\}$, it follows that $\varphi(\bar{z}^{(0)}) \leqslant \varphi(z_{\lambda_k})$. Consequently,

$$\Omega[z_{\lambda_k}] \leqslant \Omega[\bar{z}^{(0)}].$$

Thus, the sequence $\{z_{\lambda_k}\}$ belongs to the compact set of elements z for which

$$\Omega[z] \leqslant \Omega[\bar{z}^{(0)}].$$

Consequently, it has a convergence subsequence $\{z'_{\lambda_k}\}$. Let us write $\bar{z} = \lim_{k\to\infty} z'_{\lambda_k}$. Obviously, $\|\bar{z} - \bar{z}^{(0)}\| \geqslant \epsilon_0$. Since $z'_{\lambda_k} \in R_2$, we have

$$A\bar{z} = \bar{u} \quad \text{and} \quad \bar{z} \in R_1.$$

Obviously,

$$\varphi(\bar{z}) = \varphi(\bar{z}^{(0)})$$

and

$$\|\bar{z} - z^0\|^2 = \Omega[\bar{z}] \leqslant \Omega[\bar{z}^{(0)}] = \|\bar{z}^{(0)} - z^0\|^2.$$

These conditions define a unique normal solution of the problem (8.2.1)–(8.2.3). Consequently, $\bar{z} = \bar{z}^{(0)}$. But this contradicts the inequality $\|\bar{z} - \bar{z}^{(0)}\| \geqslant \epsilon_0$. This completes the proof of the theorem.

Remark. If conditions (8.2.1) and (8.2.2) are not satisfied (or if their satisfaction is difficult to determine), we can seek a quasisolution of the linear programming problem. To find it, we use the regularization method described above.

6. Suppose that the initial data A, $\bar{u}$, and $\bar{c}$ of the problem (8.2.4) are known only approximately. Instead of A, $\bar{u}$, and $\bar{c}$, we have $\tilde{A}$, $\tilde{u}$, and $\tilde{c}$ such that

$$\|\tilde{A} - A\| \leqslant \delta, \quad \|\tilde{u} - \bar{u}\| \leqslant \delta, \quad \|\bar{c} - \tilde{c}\| \leqslant \delta.$$

In this case, we can speak only of finding a solution that is close to the normal solution of the problem (8.2.4). In this subsection, we shall show that, for suitable choice of the parameters α and λ compatible with the error δ in the initial data $\tilde{A}$, $\tilde{u}$, and $\tilde{c}$, a regularized solution of the problem $z_{\alpha,\lambda}$ that minimizes the functional

$$M_\lambda^\alpha[z;\ \tilde{A}, \tilde{u}, \tilde{c}]$$

approximates with prenamed accuracy the sought normal solution $\bar{z}^{(0)}$ of the problem (8.2.4) with exact initial data A, $\bar{u}$, and $\bar{c}$ and that it is stable under small changes in A, $\bar{u}$, and $\bar{c}$.

We note first of all that the system of equations

$$\tilde{A}z = \tilde{u}$$

is consistent if and only if

$$\tilde{u} \in U_{\tilde{A}} \equiv \{u;\ u = \tilde{A}z,\ z \in R^n\}.$$

If $\tilde{u} \notin U_{\tilde{A}}$, then, when we denote by $\tilde{v}$ the orthogonal projection of $\tilde{u}$ onto $U_{\tilde{A}}$, we have

$$\|\tilde{A}z - \tilde{u}\|^2 = \|\tilde{A}z - \tilde{v}\|^2 + \|\tilde{v} - \tilde{u}\|^2.$$

It follows that

$$M_\lambda^\alpha[z;\ \tilde{A}, \tilde{u}, \tilde{c}] = M_\lambda^\alpha[z;\ \tilde{A}, \tilde{v}, \tilde{c}] + \|\tilde{v} - \tilde{u}\|^2. \quad (8.3.2)$$

The second term in the right-hand member of (8.3.2) is independent of α and λ. Consequently, the element $z_{\alpha,\lambda}$ minimizing the functional $M_\lambda^\alpha[z; \tilde{A}, \tilde{u}, \tilde{c}]$ also minimizes the functional $M_\lambda^\alpha[z; \tilde{A}, \tilde{v}, \tilde{c}]$ and vice versa. Since both these functionals are positive quadratic forms in the z_j for $j = 1, 2, \ldots, n$, the existence of an element $z_{\alpha,\lambda}$ minimizing them on R_1 is obvious.

7. Let us estimate the deviation of the regularized solution from the exact normal one.

Theorem 2. *In the problem* (8.2.1)–(8.2.3), *suppose that, instead of the exact initial data, A, $\bar{u}$ and c, we know approximate data $\tilde{A}$, $\tilde{\bar{u}}$, and $\tilde{c}$ such that*

$$\|\tilde{A}-A\|\leqslant\delta,\quad \|\tilde{\bar{u}}-\bar{u}\|\leqslant\delta,\quad \|\tilde{c}-c\|\leqslant\delta.$$

Suppose that $z_{\alpha,\lambda}$ is an element minimizing the functional

$$M_{\lambda}^{\alpha}[z;\ \tilde{A},\ \tilde{\bar{u}},\ \tilde{c}]$$

on the set R_1 and that $\bar{z}^{(0)}$ is a normal solution of the problem (8.2.1)–(8.2.3) with exact initial data. Suppose that $\alpha_0(\delta)$ and $\beta_0(\delta)$ are given nonnegative continuous increasing functions that vanish at $\delta = 0$ and satisfy the condition

$$\delta^2\leqslant\alpha_0(\delta)\,\beta_0(\delta).$$

Then, for any $\epsilon>0$, there exist $\lambda_0(\epsilon)$ and $\delta_0(\epsilon,\lambda_0)$ *(depending also on A, $\bar{u}$, c, $\alpha_0(\delta)$, and $\beta_0(\delta)$) such that the inequality*

$$\|z_{\alpha,\lambda}-\bar{z}^{(0)}\|\leqslant\epsilon$$

holds for all $\lambda\leqslant\lambda_0(\epsilon)$ and all $\delta\leqslant\delta_0(\epsilon,\lambda_0)$ and α that satisfy the double inequality

$$\frac{\delta^2}{\beta_0(\delta)}\leqslant\alpha\leqslant\alpha_0(\delta).$$

Proof. Theorem 1 of §3 tells us that, for every $\epsilon>0$, there exists a $\lambda_0(\epsilon/2)$ such that

$$\|z_{\lambda}-\bar{z}^{(0)}\|\leqslant\epsilon/2$$

for arbitrary $\lambda\leqslant\lambda_0(\epsilon/2)$ and any element z_{λ} minimizing the functional

$$\tilde{\varphi}^2(z) + \lambda\Omega[z]$$

on the set R_2. Therefore, it will be sufficient for us to show that, under the conditions of Theorem 2 of the present section, for any fixed $\lambda > 0$ there exists a $\delta_0(\epsilon, \lambda)$ such that, for all $\delta \leqslant \delta_0(\epsilon, \lambda)$ and α satisfying the double inequality

$$\frac{\delta^2}{\beta_0(\delta)} \leqslant \alpha \leqslant \alpha_0(\delta), \tag{8.3.3}$$

we have

$$\| z_{\alpha,\lambda} - z_\lambda \| \leqslant \epsilon/2. \tag{8.3.4}$$

To prove the validity of the estimate (8.3.4), let us first estimate $\Omega[\tilde{z}_{\alpha,\lambda}]$ and $\|A\tilde{z}_{\alpha,\lambda} - A\bar{z}^{(0)}\|$. Obviously,

$$\alpha\lambda\Omega[\tilde{z}_{\alpha,\lambda}] \leqslant M_\lambda^\alpha[\tilde{z}_{\alpha,\lambda}; \tilde{A}, \tilde{v}, \tilde{c}] \leqslant M_\lambda^\alpha[\bar{z}^{(0)}; \tilde{A}, \tilde{v}, \tilde{c}] =$$
$$= \|\tilde{A}\bar{z}^{(0)} - \tilde{v}\|^2 + \alpha\{\tilde{\varphi}^2(\bar{z}^{(0)}) + \lambda\Omega[\bar{z}^{(0)}]\}. \tag{8.3.5}$$

Here, $\tilde{\varphi}(z) = (\tilde{c}, z)$. Also,

$$\|\tilde{A}\bar{z}^{(0)} - \tilde{v}\| \leqslant$$
$$\leqslant \|\tilde{A}\bar{z}^{(0)} - A\bar{z}^{(0)}\| + \|A\bar{z}^{(0)} - \tilde{u}\| + \|\tilde{u} - \tilde{v}\| =$$
$$= \|\tilde{A}\bar{z}^{(0)} - A\bar{z}^{(0)}\| + \|\bar{u} - \tilde{u}\| + \|\tilde{u} - \tilde{v}\|).$$

Using the estimates $\|\tilde{A} - A\| \leqslant \delta$ and $\|\tilde{u} - \bar{u}\| \leqslant \delta$, we obtain

$$\|\tilde{A}\bar{z}^{(0)} - \tilde{v}\| \leqslant \delta\|\bar{z}^{(0)}\| + \delta + \|\tilde{u} - \tilde{v}\|. \tag{8.3.6}$$

Since $\|\tilde{u} - \tilde{v}\| \leqslant \|\tilde{u} - \tilde{A}z\|$ for an arbitrary element z, we obtain from (8.3.6)

$$\|\tilde{A}\bar{z}^{(0)} - \tilde{v}\| \leqslant \delta(1 + \|\bar{z}^{(0)}\|) + \|\tilde{u} - \tilde{A}\bar{z}^{(0)}\| \leqslant$$
$$\leqslant \delta(1 + \|\bar{z}^{(0)}\|) + \|\tilde{u} - \bar{u}\| + \|\bar{u} - \tilde{A}\bar{z}^{(0)}\| =$$

$$= \delta(1 + \|\bar{z}^{(0)}\|) + \|\tilde{u} - \bar{u}\| + \|A\bar{z}^{(0)} - \tilde{A}\bar{z}^{(0)}\| \leqslant$$

$$\leqslant \delta(1 + \|\bar{z}^{(0)}\|) + \delta + \|\tilde{A} - A\|\|\bar{z}^{(0)}\| \leqslant 2\delta(1 + \|\bar{z}^{(0)}\|).$$

Thus,

$$\|\tilde{A}\bar{z}^{(0)} - \bar{v}\| \leqslant B\delta, \tag{8.3.7}$$

where $B = 2(1 + \|\bar{z}^{(0)}\|)$. Using this estimate, we obtain from (8.3.5)

$$\alpha\lambda\Omega[\tilde{z}_{\alpha,\lambda}] \leqslant \alpha\lambda\left\{\frac{B\delta^2}{\alpha\lambda} + \frac{\tilde{\varphi}^2(\bar{z}^{(0)})}{\lambda} + \Omega[\bar{z}^{(0)}]\right\}.$$

Consequently,

$$\Omega[\tilde{z}_{\alpha,\lambda}] \leqslant \frac{B\delta^2}{\alpha\lambda} + \frac{\tilde{\varphi}^2(\bar{z}^{(0)})}{\lambda} + \Omega[\bar{z}^{(0)}].$$

Using (8.3.3), we obtain

$$\Omega[\tilde{z}_{\alpha,\lambda}] \leqslant \frac{B\cdot\beta_0(\delta)}{\lambda} + \frac{\tilde{\varphi}^2(\bar{z}^{(0)})}{\lambda} + \Omega[\bar{z}^{(0)}] = N(\delta, \lambda).$$

Since $N(\delta, \lambda)$ is an increasing function of δ, it follows that $\Omega[\tilde{z}_{\alpha,\lambda}] \leqslant N(\bar{\delta}_0, \lambda)$, where $\bar{\delta}_0$ is a fixed number.

Let us estimate $\|A\tilde{z}_{\alpha,\lambda} - A\bar{z}^{(0)}\|$. Obviously,

$$\|A\tilde{z}_{\alpha,\lambda} - A\bar{z}^{(0)}\| \leqslant$$
$$\leqslant \|A\tilde{z}_{\alpha,\lambda} - \tilde{A}\tilde{z}_{\alpha,\lambda}\| + \|\tilde{A}\tilde{z}_{\alpha,\lambda} - \tilde{v}\| + \|\tilde{v} - A\bar{z}^{(0)}\| \leqslant$$
$$\leqslant \|A\tilde{z}_{\alpha,\lambda} - \tilde{A}\tilde{z}_{\alpha,\lambda}\| + \{M_\lambda^\alpha[\tilde{z}_{\alpha,\lambda}; \tilde{A}, \tilde{v}, \tilde{c}]\}^{1/2} + \|\tilde{v} - A\bar{z}^{(0)}\|.$$

Using (8.3.5), (8.3.7), and (8.3.3), we obtain

$$\|A\tilde{z}_{\alpha,\lambda} - A\bar{z}^{(0)}\| \leqslant \delta\|\tilde{z}_{\alpha,\lambda}\| +$$
$$+ \sqrt{B\delta^2 + \alpha_0(\delta)\{\tilde{\varphi}^2(\bar{z}^{(0)}) + \lambda\Omega[\bar{z}^{(0)}]\}} + \|\tilde{v} - A\bar{z}^{(0)}\|.$$

Since $A\bar{z}^{(0)}=\bar{u}$ and $\|\tilde{u}-\tilde{v}\|\leqslant\|\tilde{u}-\tilde{A}z\|$ for every z, we have

$$\|\tilde{v}-A\bar{z}^{(0)}\|\leqslant\|\tilde{u}-\tilde{v}\|+\|\tilde{u}-\bar{u}\|\leqslant$$
$$\leqslant\|\tilde{u}-\tilde{A}\bar{z}^{(0)}\|+\delta\leqslant\|\tilde{u}-\bar{u}\|+\|A\bar{z}^{(0)}-\tilde{A}\bar{z}^{(0)}\|+\delta\leqslant$$
$$\leqslant 2\delta+\|\tilde{A}-A\|\|\bar{z}^{(0)}\|\leqslant\delta(2+\|\bar{z}^{(0)}\|).$$

Using this estimate, we obtain

$$\|A\tilde{z}_{\alpha,\lambda}-A\bar{z}^{(0)}\|\leqslant\delta(2+\|\bar{z}^{(0)}\|+\|\tilde{z}_{\alpha,\lambda}\|)+$$
$$+\sqrt{B\delta^2+\alpha_0(\delta)\{\tilde{\varphi}^2(\bar{z}^{(0)})+\lambda\Omega[\bar{z}^{(0)}]\}}=\eta(\delta).$$

Obviously, $\eta(\delta)\to\infty$ as $\delta\to 0$.

Let us now show that, for arbitrary $\epsilon>0$, there exists a $\delta_0 = \delta_0(\epsilon, \lambda)$ such that the inequality

$$\|\tilde{z}_{\alpha,\lambda}-z_\lambda\|\leqslant\epsilon/2$$

holds for $\delta\leqslant\delta_0(\epsilon, \lambda)$ and all α satisfying (8.3.3). Let us suppose that no such $\delta_0(\epsilon, \lambda)$ exists. This means that there exist an $\epsilon_0>0$ and sequences $\{\tilde{A}_k\}$, $\{\tilde{u}_k\}$, $\{\tilde{c}_k\}$, and $\{\delta_k\}$ that converge (with respect to the corresponding norms) to A, $\bar{u}$, c, and 0 respectively such that the inequality

$$\|\tilde{z}_{\alpha_k,\lambda}-z_\lambda\|\geqslant\epsilon_0/2$$

holds for all α_k satisfying the conditions

$$\frac{\delta_k^2}{\beta_0(\delta_k)}\leqslant\alpha_k\leqslant\alpha_0(\delta_k).$$

However,

$$\Omega[\tilde{z}_{\alpha_k,\lambda}]\leqslant N(\bar{\delta}_0, \lambda).$$

Consequently, the sequence $\{\tilde{z}_{\alpha_k,\lambda}\}$ belongs to a compact set and hence has a convergent subsequence $\{\widetilde{z}'_{\alpha_k,\lambda}\}$. Let us define

$$\widetilde{z}'_\lambda = \lim_{k\to\infty} \widetilde{z}'_{\alpha_k,\lambda}.$$

Obviously, $\widetilde{z}'_\lambda \in R_1$. Since

$$\| A\widetilde{z}'_{\alpha_k,\lambda} - A\bar{z}^{(0)} \| \leqslant \eta(\delta_k)$$

and $\eta(\delta_k) \to 0$ as $k \to \infty$, passage to the limit yields

$$\| A\widetilde{z}'_\lambda - A\bar{z}^{(0)} \| = 0,$$

so that $A\widetilde{z}'_\lambda = A\bar{z}^{(0)} = \bar{u}$. This means that $\widetilde{z}'_\lambda \in R_2$.

Let us estimate $\varphi^2(\widetilde{z}'_{\alpha_k,\lambda}) + \lambda\Omega[\widetilde{z}'_{\alpha_k,\lambda}]$. If we write $\widetilde{\varphi}_k(z) = (\widetilde{c}_k, z)$, we have

$$\begin{aligned}
&\alpha_k \{\varphi^2(\widetilde{z}'_{\alpha_k,\lambda}) + \lambda\Omega[\widetilde{z}'_{\alpha_k,\lambda}]\} \leqslant \\
&\quad \leqslant \alpha_k \{\varphi^2(\widetilde{z}'_{\alpha_k,\lambda}) - \widetilde{\varphi}^2_k(\widetilde{z}'_{\alpha_k,\lambda}) + \widetilde{\varphi}^2_k(\widetilde{z}'_{\alpha_k,\lambda}) + \lambda\Omega[\widetilde{z}'_{\alpha_k,\lambda}]\} \leqslant \\
&\leqslant \alpha_k \{\widetilde{\varphi}^2_k(\widetilde{z}'_{\alpha_k,\lambda}) + \lambda\Omega[\widetilde{z}'_{\alpha_k,\lambda}]\} + \alpha_k \| \varphi^2(\widetilde{z}'_{\alpha_k,\lambda}) - \widetilde{\varphi}^2_k(\widetilde{z}'_{\alpha_k,\lambda}) \| = \\
&\qquad = \alpha_k \{\widetilde{\varphi}^2_k(\widetilde{z}'_{\alpha_k,\lambda}) + \lambda\Omega[\widetilde{z}'_{\alpha_k,\lambda}]\} + \\
&\quad + \alpha_k \| \varphi(\widetilde{z}'_{\alpha_k,\lambda}) - \widetilde{\varphi}_k(\widetilde{z}'_{\alpha_k,\lambda}) \| \cdot \| \varphi(\widetilde{z}'_{\alpha_k,\lambda}) + \widetilde{\varphi}_k(\widetilde{z}'_{\alpha_k,\lambda}) \| \leqslant \\
&\qquad \leqslant \alpha_k \{\widetilde{\varphi}^2_k(\widetilde{z}'_{\alpha_k,\lambda}) + \lambda\Omega[\widetilde{z}'_{\alpha_k,\lambda}]\} + \\
&\quad + \alpha_k \| (\widetilde{c}_k, \widetilde{z}'_{\alpha_k,\lambda}) - (c, \widetilde{z}'_{\alpha_k,\lambda}) \| (\| (c, \widetilde{z}'_{\alpha_k,\lambda}) \| + \| (\widetilde{c}_k, \widetilde{z}'_{\alpha_k,\lambda}) \|) \leqslant \\
&\qquad \leqslant \alpha_k \{\widetilde{\varphi}^2_k(\widetilde{z}'_{\alpha_k,\lambda}) + \lambda\Omega[\widetilde{z}'_{\alpha_k,\lambda}]\} + \\
&\qquad + \alpha_k \| \widetilde{c}_k - c \| \| \widetilde{z}'_{\alpha_k,\lambda} \| (\| c \| + \| \widetilde{c}_k \|) \| \widetilde{z}'_{\alpha_k,\lambda} \| \leqslant \\
&\leqslant \alpha_k \{\widetilde{\varphi}^2_k(\widetilde{z}'_{\alpha_k,\lambda}) + \lambda\Omega[\widetilde{z}'_{\alpha_k,\lambda}]\} + \alpha_k\delta_k \| \widetilde{z}'_{\alpha_k,\lambda} \|^2 (\| c \| + \| \widetilde{c}_k \|) \leqslant \\
&\qquad\qquad \leqslant M^{\alpha_k}_\lambda [z_\lambda, \widetilde{A}_k, \widetilde{u}_k, \widetilde{c}_k] + O(\alpha_k \cdot \delta_k).
\end{aligned}$$

Thus,

$$\alpha_k \{\varphi^2 (\tilde{z}'_{\alpha_k,\lambda}) + \lambda\Omega [\tilde{z}'_{\alpha_k,\lambda}]\} \leqslant$$

$$\leqslant M_k^{\alpha\lambda} [z_\lambda, \tilde{A}_k, \tilde{u}_k, \tilde{c}_k] + O (\alpha_k \cdot \delta_k). \quad (8.3.8)$$

Here, z_k is an element minimizing the functional $\varphi^2(z) + \lambda\Omega[z]$ on the set R_2.

In a completely analogous manner, we obtain

$$\alpha_k \{\tilde{\varphi}_k^2 (z_\lambda) + \lambda\Omega [z_\lambda]\} \leqslant$$

$$\leqslant \alpha_k \{\varphi^2 (z_\lambda) + \lambda\Omega [z_\lambda]\} + \alpha_k\delta_k \| z_\lambda \|^2 (\| c \| + \| \tilde{c}_k \|). \quad (8.3.9)$$

Let us estimate

$$M_\lambda^{\alpha_k} [z_\lambda; \tilde{A}_k, \tilde{u}_k, \tilde{c}_k] = \| \tilde{A}_k z_\lambda - \acute{u}_k \|^2 + \alpha_k \{\tilde{\varphi}_k^2 (z_\lambda) + \lambda\Omega [z_\lambda]\}.$$

Since $Az_\lambda = \bar{u}$ and $\|z_\lambda\| \leqslant \|\bar{z}^{(0)}\|$, we have

$$\tilde{A}_k z_\lambda - \tilde{u}_k \| \leqslant \| \tilde{A}_k z_\lambda - A z_\lambda \| + \| \tilde{u}_k - \bar{u} \| \leqslant$$

$$\leqslant \| \tilde{A}_k - A \| \| z_\lambda \| + \delta_k \leqslant \delta_k (1 + \| z_\lambda \|) \leqslant \delta_k (1 + \| \bar{z}^{(0)} \|).$$

Using (8.3.9), we obtain

$$M_\lambda^{\alpha_k} [z_\lambda, \tilde{A}_k, \tilde{u}_k, \tilde{c}_k] \leqslant \delta_k^2 (1 + \| \bar{z}^{(0)} \|)^2 +$$

$$+ \alpha_k \{\varphi^2 (z_\lambda) + \lambda\Omega [z_\lambda]\} + \alpha_k \cdot \delta_k (\| c \| + \| \tilde{c}_k \|) \| z_\lambda \|^2.$$

It follows from this estimate and (8.3.8) that

$$\varphi^2 (\tilde{z}'_{\alpha_k,\lambda}) + \lambda\Omega [\tilde{z}'_{\alpha_k,\lambda}] \leqslant \varphi^2 (z_\lambda) + \lambda\Omega [z_\lambda] + O (\delta_k) + O (\delta_k^2/\alpha_k).$$

Taking the limit as $k \to \infty$, we obtain

$$\varphi^2 (\tilde{z}_\lambda) + \lambda\Omega [\tilde{z}_\lambda] \leqslant \varphi^2 (z_\lambda) + \lambda\Omega [z_\lambda].$$

Since the element z_λ minimizes the functional $\varphi^2(z) + \lambda\Omega[z]$, we have $\tilde{z}_\lambda = z_\lambda$. Consequently, for sufficiently large k, say for $k \geqslant k_0(\epsilon_0)$, we have $\|\tilde{z}'_{\alpha_k, \lambda_1} - z_\lambda\| < \epsilon_0/2$, which contradicts the assumption. This completes the proof of the theorem.

Remark. We took $\Omega[z] = \|z - z^0\|^2$ but actually used only the fact that the set of elements z for which $\Omega[z] \leqslant d$ is compact for arbitrary $d > 0$. All the results remain valid if we take for $\Omega[z]$ any positive-definite quadratic form

$$\Omega[z] = \sum_{i,j} p_{ij}(z_i - z_i^0)(z_j - z_j^0),$$

modifying appropriately the definition of a normal solution.

BIBLIOGRAPHY

1. L. Aleksandrov, A regularizational computational process for analyzing exponential dependence, *Zhurnal vychislitel'noy matematiki i matematicheskoy fiziki*, **10**, 5 (1970).
2. B. Aliyev, Two approaches to the difference method of solving Neumann's problem in a rectangular region. *Zhurnal vychislitel'noy matematiki i matematicheskoy fiziki*, **12**, 1 (1972).
3. B. Aliyev, Regularizing algorithms for a stable normal solution of an equation of the second kind on a spectrum, *Zhurnal vychislitel'noy matematiki i matematicheskoy fiziki*, **10**, 3 (1970).
4. Yu. B. Anikonov, Operator equations of the first kind, *Doklady Akad. nauk SSSR,* **207**, 2, (1972).
5. Yu. T. Antokhin, An analytical approach to the problem of equations of the first kind, *Doklady Akad. nauk SSSR,* **167**, 4 (1966).
6. Yu. T. Antokhin, Ill-posed problems in a Hilbert space and stable methods of solving them, *Differentsial'nyye Uravneniya,* **3**, 7 (1967).
7. Yu. T. Antokhin, Ill-posed problems for equations of the convolution type, *Differentisial'nyye Uravneniya,* **4**, 9 (1968).

8. V. Ya. Arsenin, Discontinuous solutions of equations of the first kind, *Zhurnal vychislitel'noy matematiki i matematicheskoy fiziki,* **5**, 5 (1965).
9. V. Ya. Arsenin and V. V. Ivanov, Solution of certain integral equations of the first kind of the convolution type by the regularization method. *Zhurnal vychislitel'noy matematiki i matematicheskoy fiziki,* **8**, 2 (1968).
10. V. Ya. Arsenin and V. V. Ivanov, The influence of *p*th-order regularization, *Zhurnal vychislitel'noy matematiki i matematicheskoy fiziki,* **8**, 3 (1968).
11. V. Ya. Arsenin and V. V. Ivanov, Optimal regularization, *Doklady Akad. nauk SSSR,* **182**, 1 (1968).
12. V. Ya. Arsenin, Optimal summation of Fourier series with approximate coefficients, *Doklady Akad. Nauk SSSR,*, **183**, 2 (1968).
13. V. Ya. Arsenin and T. I. Savelova, Application of the regularization method to integral equations of the first kind of the convolution type, *Zhurnal vychislitel'noy matematiki i matematicheskoy fiziki,* **9**, 6 (1969).
14. V. Ya. Arsenin, A method of approximate solution of integral equations of the first kind of the convolution type *Trudy* (Proceedings) *Mat. Inst. Akad. Nauk SSSR,* **133** (1973).
15. V. Ya. Arsenin, *O Metodakh resheniya nekorrektno postavlennykh zadach* (Methods of solving ill-posed problems), Rotoprint, Moscow Institute of Engineering Physics (MIFI), Moscow, 1973.
16. A. B. Bakushinskiy, A general procedure for constructing regularizing algorithms for a linear ill-posed equation in a Hilbert space, *Zhurnal vychislitel'noy matematiki i matematicheskoy fiziki,* **7**, 3 (1968).
17. A. B. Bakushinskiy and V. N. Strakhov, The solution of certain integral equations of the first kind by the method of approximations, *Zhurnal vychislitel'noy matematiki i matematicheskoy fiziki,* **8**, 1 (1968).
18. A. B. Bakushinskiy, Algorithms for regularization in the case

of linear equations with unbounded operators, *Doklady Akad. Nauk SSSR,* **183**, 1 (1968).

19. A. B. Bakushinskiy, The problem of constructing linear regularizing algorithms in Banach spaces, *Zhurnal vychislitel'-noy matematiki i matematicheskoy fiziki,* **13**, 1 (1973).
20. B. M. Budak and V. N. Vasil'yeva, The solution of the inverse Stefan problem, *Zhurnal vychislitel'noy matematiki i matematicheskoy fiziki,* **13**, 1, 4 (1973).
21. V. V. Vasin, Regularization of nonlinear partial differential equations, *Differentsial'nyye Uravneniya,* **4**, 12 (1968).
22. V. V. Vasin and V. P. Tanana, Approximate solution of operator equations of the first kind, *Matematicheskiye zapiski*, Gorky University of the Urals, **4**, 6 (1968).
23. V. V. Vasin, The connection between different variational methods of approximate solution of ill-posed problems, *Matematicheskiye zametki,* **7**, 3 (1970).
24. V. V. Vasin, Stable computation of a derivative, *Zhurnal vichislitel'noy matematiki i matematicheskoy fiziki,* **13**, 6 (1973).
25. V. A. Vinokurov, The concept of regularizability of discontinuous mappings, *Zhurnal vychislitel'noy matematiki i matematicheskoy fiziki,* **11**, 5 (1971).
26. V. A. Vinokurov, The error in approximate solution of linear problems, *Zhurnal vychislitel'noy matematiki i matematiche-skoy fiziki,* **12**, 3 (1972).
27. V. A. Vinokurov, General properties of the error of approximation of the solution of linear functional equations, *Zhurnal vychislitel'noy matematiki i matematicheskoy fiziki,* **11**, 1 (1971).
28. V. A. Vinokurov, Two notes regarding the choice of regularization parameter, *Zhurnal vychislitel'noy matematiki i matematicheskoy fiziki,* **12**, 2 (1972).
29. V. V. Voyevodin, A regularization method, *Zhurnal vychislitel'noy matematiki i matematicheskoy fiziki,* **9**, 3 (1969).
30. M. K. Gavurin and V. M. Ryabov, Application of Chebyshev polynomials to regularization of ill-posed and poorly

conditioned equations in a Hilbert space, *Zhurnal vychislitel'-noy matematiki i matematicheskoy fiziki,* **13**, 6 (1973).
31. V. Ya. Galkin, Calculation of the controlling function upon the lifting of a rocket probe to the maximum altitude, in the collection *Vychislitel'naya matematika i programmirovaniye* (Computational mathematics and programming), **XII**, Moscow State Univ. Press, 1969.
32. V. Ya. Galkin, The problem of relay control in vertical lifting of a rocket, in the collection *Vychislitel'naya matematika i programmirovaniye* (Computational mathematics and programming), **XII**, Moscow State University Press, 1969.
33. V. Ya. Galkin, Problems of processing and interpretation of the results of certain experiments in nuclear physics, author's abstract of candidate's dissertation, Moscow State University, 1972.
34. T. M. Geyman, P. N. Zaikin, V. A. Maslennikov, and N. N. Sedov, Problems in quantitative electronic microscopy, in the collection *Vychislitel'naya matematika i programmirovaniye* (Computational mathematics and programming), **XIV**, Moscow State University Press, 1970.
35. V. B. Glasko, A. N. Tikhonov, and A. V. Tikhonravov, Synthesis of multilayer coverings, *Zhurnal vychislitel'noy matematiki i matematicheskoy fiziki,* **14**, 1 (1974).
36. V. B. Glasko, V. V. Kravtsov, and G. N. Kravtsova, A reverse gravimetry problem *Vestnik* (Herald), Moscow State University, 2 (1970).
37. V. B. Glasko, Uniqueness of solution of certain inverse seismology problems, *Zhurnal vychislitel'noy matematiki i matematicheskoy fiziki,* **10**, 6 (1970).
38. V. B. Glasko, A. Kh. Ostromogil'skiy, and V. G. Filatov, Restoration of the depth and form of a contact surface by regularization, *Zhurnal vychislitel'noy matematiki i matematicheskoy fiziki,* **10**, 5 (1970).
39. V. B. Glasko, Use of the regularization method to solve the problem of thermal probing of the atmosphere, *Fizika atmosfery i okeana,* **IV**, 3 (1968).

40. V. B. Glasko, B. A. Volodin, Ye. A. Mudretsov, and N. Yu. Nefedova, Solution of the inverse gravitational prospecting problem for a contact surface by the regularization method, *Fizika Zemli,* **5** (1972).
41. V. B. Glasko, Uniqueness of determination of the structure of the crust of the earth from Rayleigh surface waves, *Zhurnal matematiki i matematicheskoy fiziki,* **11**, 6 (1971).
42. V. B. Glasko, Some mathematical questions of interpreting geophysical observations, author's abstract of doctor's dissertation, Moscow State University, 1972.
43. A. V. Goncharskiy and A. G. Yagola, Uniform approximation of monotonic solutions of ill-posed problems, *Doklady Akad. Nauk SSSR,* **184**, 4 (1969).
44. A. V. Goncharskiy, A. S. Leonov, and A. G. Yagola, Estimates of the speed of convergence of regularized approximations for equations of the convolution type, *Zhurnal vychislitel'noy matematiki i matematicheskoy fiziki,* **12**, 3 (1972).
45. A. V. Goncharskiy, A. S. Leonov, and A. G. Yagola, Solution of two-dimensional integral equations of the first kind with kernel depending on the difference between its arguments, *Zhurnal vychislitel'noy matematiki i matematicheskoy fiziki,* **11**, 5 (1971).
46. A. V. Goncharskiy, A. S. Leonov, and A. G. Yagola, A regularizing algorithm for ill-posed problems with approximately given operator, *Zhurnal vychislitel'noy matematiki i matematicheskoy fiziki,* **12**, 6 (1972).
47. A. V. Goncharskiy, A. S. Leonov, and A. G. Yagola, A generalized discrepancy principle, *Zhurnal vychislitel'noy matematiki i matematicheskoy fiziki,* **13**, 2 (1973).
48. A. V. Goncharskiy, A. S. Leonov, and A. G. Yagola, A finite-difference approximation of linear ill-posed problems, *Zhurnal vychislitel'noy matematiki i matematicheskoy fiziki,* **14**, 1 (1974).
49. A. V. Goncharskiy, A. S. Leonov, and A. G. Yagola, A discrepancy Principle in the solution of nonlinear ill-posed problems, *Doklady Akad. Nauk USSR,* **214**, 3 (1974).

50. V. I. Gordonova and V. A. Morozov, Numerical Algorithms for choosing a parameter in the regularization method, *Zhurnal vychislitel'noy matematiki i matematicheskoy fiziki,* **13**, 3 (1973).
51. V. B. Demidovich, Restoration of a function and its derivatives from experimental information, in the collection *Vychislitel'naya matematika i programmirovaniye,* **VIII**, Moscow State University Press, 1967.
52. A. M. Denisov, Approximations of quasisolutions of a Fredholm equation of the first kind with kernel of a special form, *Zhurnal vychislitel'noy matematiki i matematicheskoy fiziki,* **11**, 5 (1971) and **12**, 6 (1972).
53. A. M. Denisov, Approximation of quasisolutions of certain integral equations of the first kind, *Zhurnal vychislitel'noy matematiki i matematicheskoy fiziki,* **14**, 1 (1974).
54. R. Denchev, Tikhonov's regularization method for weak solutions of boundary-value problems, *Zhurnal vychislitel'noy matematiki i matematicheskoy fiziki,* **9**, 2 (1969).
55. V. P. Didenko and N. N. Kozlov, Regularization of certain ill-posed problems in engineering cybernetics, *Doklady Akad. Nauk SSSR,* **214**, 3 (1974).
56. T. F. Dolgopolova, Finite-dimensional regularization in the case of numerical differentiation of periodic functions, *Matematicheskiye zapiski,* Gorky University of the Urals, **7**, 4(1970).
57. T. F. Dolgopolova and V. K. Ivanov, Numerical differentiation, *Zhurnal vychislitel'noy matematiki i matematicheskoy fiziki,* **6**, 3 (1966).
58. I. N. Dombrovskaya, Linear operator equations of the first kind, *Izvestiya vuzov, Matematika,* **2** (1964).
59. I. N. Dombrovskaya, Solution of ill-posed problems in a Hilbert space, *Matematicheskiye zapiski*, Gorky University of the Urals, **4**, 4 (1964).
60. I. N. Dombrovskaya and V. K. Ivanov, The theory of certain linear equations in abstract spaces, *Sibirskiy matematicheskiy zhurnal,* **VI**, 3 (1965).
61. I. N. Dombrovskaya, Equations of the first kind with closed operator, *Izvestiya vuzov, Matematika,* **6** (1967).

62. Ye. L. Zhukovskiy and R. Sh. Liptser, A recursion method of calculating normal solutions of linear algebraic systems of equations, *Zhurnal vychislitel'noy matematiki i matematicheskoy fiziki,* **12**, 4 (1972).
63. Ye. L. Zhukovskiy and V. A. Morozov, Successive Bayesian regularization of algebraic systems of equations, *Zhurnal vychislitel'noy matematiki i matematicheskoy fiziki,* **12**, 2 (1972).
64. Ye. L. Zhukovskiy, Statistical regularization of algebraic systems of equations, *Zhurnal vychislitel'noy matematiki i matematicheskoy fiziki,* **12**, 1 (1972).
65. P. N. Zaikin, Numerical solution of the inverse problem of operational calculus in the real domain, *Zhurnal vychislitel'noy matematiki i matematicheskoy fiziki,* **8**, 2 (1968).
66. P. N. Zaikin, A system of continuous automatic processing of experimental results from investigation of sections of photonuclear reactions, Author's abstract of candidate's dissertation, Moscow State University, 1968.
67. P. N. Zaikin and A. S. Mechenov, Questions regarding numerical solution of integral equations of the first kind by the regularization method, *Otchet vychislitel'nogo tsentra* (Report of the computational center), Moscow State University, No. 144–TZ, 1971 (rotaprint).
68. V. K. Ivanov, Unstable linear problems with multiple-valued operators, *Sibirskiy matematicheskiy zhurnal,* **XI**, 5 (1970).
69. V. K. Ivanov, The inverse potential problem for a body close to a given one, *Izvestiya Akad. Nauk SSSR*, Ser. matem., **20**, 6 (1956).
70. V. K. Ivanov, Stability of the inverse logarithmic potential problem, *Izvestiya vuzov, Matematika,* **4** (1958).
71. V. K. Ivanov, Ill-posed linear problem, *Doklady Akad. Nauk SSSR,* **145**, 2 (1962).
72. V. K. Ivanov, On ill-posed problems, *Matematicheskiy sbornik,* **61**, 2 (1963).
73. V. K. Ivanov, Ill-posed problems in topological spaces, *Sibirskiy matematicheskiy zhurnal,* **X**, 5 (1969).

74. V. K. Ivanov, A type of ill-posed linear equation in topological vector spaces, *Sibirskiy matematicheskiy zhurnal*, **VI**, 4 (1965).
75. V. K. Ivanov, The Cauchy problem for Laplace's problem in an infinite strip, *Differentsial'nyye Uravneniya,* **1**, 1 (1965).
76. V. K. Ivanov, Regularization of unstable problems, *Sibirskiy matematicheskiy zhurnal,* **VII**, 3 (1966).
77. V. K. Ivanov, Approximate solution of operator equations of the first kind, *Zhurnal vychislitel'noy matematikii i matematicheskoy fiziki,* **6**, 6 (1966).
78. V. K. Ivanov and T. I. Korolyuk, A problem of numerical analytic continuation of harmonic functions, *Matematicheskiye zapiski*, Gorky University of the Urals, **5**, 4 (1966).
79. V. K. Ivanov, Fredholm integral equations of the first kind, *Differentsial'nyye Uravneniya,* **3**, 3 (1967).
80. V. K. Ivanov and T. I. Korolyuk, Estimation of the error in the solution of ill-posed linear problems, *Zhurnal vychislitel'noy matematiki i matematicheskoy fiziki,* **9**, 1 (1969).
81. I. I. Iyevlev, Approximate solution of equations of the first kind, *Zhurnal vychislitel'noy matematiki i matematicheskoy fiziki,* **13**, 4 (1973).
82. V. G. Karmanov, Estimates of the convergence of iterational minimization methods, *Zhurnal vychislitel'noy matematiki i matematicheskoy fiziki,* **14**, 1 (1974).
83. A. V. Knyazev, Conditions for proper posing of nonlinear integral equations with kernel dependent on the difference between the variables, *Zhurnal vychislitel'noy matematiki i matematicheskoy fiziki,* **10**, 4 (1970).
84. T. I. Korolyuk, The Cauchy problem for Laplace's equation, *Izvestiya vuzov, Matematika,* **3** (1973).
85. L. F. Korkina, The solution of operator equations of the first kind in Hilbert spaces, *Izvestiya vuzov, Matematika,* **7** (1967).
86. L. F. Korkina, Regularization of operator equations of the first kind, *Izvestiya vuzov, Matematika,* **8** (1969).
87. Ye. L. Kosarev, Numerical solution of Abel's integral

equation, *Zhurnal vychislitel'noy matematiki i matematicheskoy fiziki,* **13**, 6 (1973).

88. S. G. Kreyn and O. I. Prozorovskaya, Approximate methods of solving ill-posed problems, *Zhurnal vychislitel'noy matematiki i matematicheskoy fiziki,* **3**, 1 (1963).
89. S. G. Kreyn, Classes of proper posing for certain boundary-value problems, *Doklady Akad. Nauk SSSR,* **114**, 6 (1957).
90. N. M. Krukovskiy, On Tikhonov-stable summation of Fourier series with disturbed coefficients by certain regular methods, *Vestnik,* Moscow State University, Ser. I. Matematika, mekhanika, **3** (1973).
91. A. V. Kryanev, Solution of ill-posed problems by the method of successive approximations, *Doklady Akad. Nauk USSR,* **210**, 1 (1973).
92. A. V. Kryanev, An iterational method of solving ill-posed problems, *Zhurnal vychislitel'noy matematiki i matematicheskoy fiziki*, **14**, 1 (1974).
93. R. Courant and D. Hilbert, *Methods of Mathematical Physics,* II, *Partial Differential Equations,* New York, Interscience, 1962.
94. M. M. Lavrent'yev, The Cauchy problem for Laplace's equation, *Izvestiya Akad. Nauk SSSR*, Ser. matem., **20**, (1956).
95. M. M. Lavrent'yev, The inverse problem in potential theory, *Doklady Akad. Nauk SSSR,* **106**, 3 (1956).
96. M. M. Lavrent'yev, The Cauchy problem for linear elliptic equations, *Doklady Akad. Nauk SSSR,* **112**, 2 (1957).
97. M. M. Lavrent'yev, Integral equations of the first kind, *Doklady Akad. Nauk SSSR,* **127**, 1 (1959).
98. M. M. Lavrent'yev, Integral equations of the first kind, *Doklady Akad. Nauk SSSR,* **133**, 2 (1960).
99. M. M. Lavrent'yev, *Some Improperly Posed Problems of Mathematical Physics*, Springer-Verlag, Berlin, Heidelberg, and New York, 1967 (translation of *O nekotorykh nekorrektnykh zadachakh matematicheskoy fiziki*).
100. M. M. Lavrent'yev and V. G. Vasil'yev, Certain ill-posed problems of mathematical physics, *Sibirskiy matematicheskiy zhurnal,* **VII**, 3 (1966).

101. M. M. Lavrent'yev, An inverse problem for the wave equation, *Doklady Akad. Nauk SSSR,* **157**, 3, 1964.
102. M. M. Lavrent'yev, A class of inverse problems for differential equations, *Doklady Akad. Nauk SSSR,* **160**, 1 (1965).
103. M. M. Lavrent'yev, V. G. Romanov, and V. G. Vasil'yev, *Mnogomernyye obratnyye zadachi dlya differentsial'nykh uravneniy* (Multidimensional inverse problems for differential equations), Novosibirsk, Nauka press, 1969.
104. R. Lattès and J.-L. Lions, *Methods of Quasireversibility: Applications to Partial Differential Equations*, American Elsevier, New York, 1969 (translation of *Méthode de quasi-réversibilité et applications*).
105. V. L. Lebedev, On the solution of compact sets of certain restoration problems, *Zhurnal vychislitel'noy matematiki i matematicheskoy fiziki,* **6**, 6 (1966).
106. O. A. Liskovets, Ill-posed problems with a closed irreversible operator, *Differentsial'nyye Uravneniya,* **3**, 4 (1967).
107. O. A. Liskovets, Regularization of linear equations in Banach spaces, *Differentsial'nyye Uravneniya,* **4**, 6 (1968).
108. O. A. Liskovets, A regularization method for nonlinear problems with a closed operator, *Sibirskiy matematicheskiy zhurnal,* **XII**, 6 (1971).
109. O. A. Liskovets, The method of ϵ-quasisolutions for equations of the first kind, *Differentsial'nyye Uravneniya,* **9**, 10 (1973).
110. G. I. Marchuk and S. A. Atanbayev, Certain questions of global regularization, *Doklady Akad. Nauk SSSR,* **190**, 3 (1970).
111. G. I. Marchuk, The posing of certain inverse problems, *Doklady Akad. Nauk SSSR,* **156**, 3 (1964).
112. G. I. Marchuk and V. G. Vasiliyev, Approximate solution of operator equations of the first kind, *Doklady Akad. Nauk SSSR,* **195**, 4 (1970).
113. V. P. Maslov, Regularization of ill-posed problems for singular integral equations, *Doklady Akad. Nauk SSSR,* **176**, 5 (1967).

114. I. V. Mel'nikova, The solution of equations of the first kind with closed multiple-valued operator, *Izvestiya vuzov, Matematika*, **12** (1971).
115. V. A. Morozov, Regularization of ill-posed problems and the choice of regularization parameter, *Zhurnal vychislitel'noy matematiki i matematicheskoy fiziki,* **6**, 1 (1966).
116. V. A. Morozov, The choice of parameter in solving functional equations by the regularization method, *Doklady Akad. Nauk SSSR,* **175**, 6 (1967).
117. V. A. Morozov, Regularization families of operators, in the collection *Vychislitel'naya matematika i programmirovaniye* (Computational mathematics and programming), **VIII**, Moscow University Press, 1967.
118. V. A. Morozov, The principle of discrepancy in the solution of inconsistent equations by Tikhonov's regularization method, *Zhurnal vychislitel'noy matematiki i matematicheskoy fiziki*, **13**, 5 (1973).
119. V. A. Morozov, Calculation of lower bounds of functionals from approximate information, *Zhurnal vychislitel'noy matematiki i matematicheskoy fiziki,* **13**, 4 (1973).
120. V. A. Morozov, The principle of optimal discrepancy with approximate solution of equations involving nonlinear operators, *Zhurnal vychislitel'noy matematiki i matematicheskoy fiziki*, **14**, 2 (1974).
121. V. A. Morozov, Pseudosolutions, *Zhurnal vychislitel'noy matematiki i matematicheskoy fiziki,* **9**, 6 (1969).
122. V. A. Morozov, *Lineynyye i nelineynyye nekorrektnyye zadachi. Itogi nauki i tekhniki. Matematicheskiy analiz* (Linear and nonlinear ill-posed problems, contributions to science and technology, mathematical analysis), 11, Moscow, VINITI Press, 1973.
123. V. A. Morozov, Solution of ill-posed problems with nonlinear unbounded operator by the regularization method, *Differentisial'nyye Uravneniya,* **6**, 8 (1970).
124. M. V. Murav'eva. Optimality and limiting properties of a Bayesian solution of a system of linear algebraic equations,

Zhurnal vychislitel'noy matematiki i matematicheskoy fiziki, **13**, 4 (1973).
125. P. S. Novikov, Uniqueness of the inverse potential theory problem, *Doklady Akad. Nauk SSSR,* **18**, 3 (1938).
126. A. Kh. Ostromogil'skiy, Uniqueness of the solution of certain inverse problems, *Zhurnal vychislitel'noy matematiki i matematicheskoy fiziki,* **11**, 1 (1971).
127. D. Ye. Okhotsimskiy, The theory of motion of rockets, *Prikladnaya matematika i mekhanika,* **10**, 2 (1946).
128. A. P. Petrov and A. V. Khovanskiy, Estimation of the error in solution of linear problems when there are errors in the operators and right-hand members, *Zhurnal vychislitel'noy matematiki i matematicheskoy fiziki,* **14**, 2 (1974).
129. V. V. Petrov and A. S. Uskov, Informational aspects of the regularization problem, *Doklady Akad. Nauk SSSR,* **195**, 4 (1970).
130. A. Ya. Perel'man and V. A. Punina, Application of Mellin's convolution to the solution of integral equations of the first kind with kernel depending on a product, *Zhurnal vychislitel'noy matematiki i matematicheskoy fiziki,* **9**, 3 (1969).
131. A. I. Prilepko, The external inverse problem of the volumetric potential of variable density for a body close to a given one, *Doklady Akad. Nauk SSSR,* **185**, 1 (1969).
132. A. I. Prilepko, Interior inverse problems in potential theory, *Doklady Akad. Nauk SSSR,* **182**, 3 (1968).
133. A. I. Prilepko, Inverse contact problems of generalized magnetic potentials, *Doklady Akad. Nauk SSSR,* **181**, 5 (1968).
134. A. I. Prilepko, Uniqueness of determination of the form and density of a body in inverse potential theory problems, *Doklady Akad. Nauk SSSR,* **193**, 2 (1970).
135. A. I. Prilepko, Uniqueness of determination of the form of a body from the values of the external potential, *Doklady Akad. Nauk SSSR,* **160**, 1 (1965).

136. A. I. Prilepko, Uniqueness of the solution of an inverse problem in the form of an integral equation of the first kind, *Doklady Akad. Nauk SSSR,* **167**, 4 (1966).
137. A. I. Prilepko, Existence of solutions of inverse problems in potential theory, *Doklady Akad. Nauk SSSR,* **199**, 1 (1971).
138. A. I. Prilepko, Interior inverse problems regarding generalized potentials, *Sibirskiy matematicheskiy zhurnal,* **12**, 3 (1971).
139. A. I. Prilepko, Inverse problem in potential theory, *Differentsil'nyye Uravneniya,* **3**, 1 (1967).
140. A. I. Prilepko, The interior inverse problem of metaharmonic potential for a body close to a given one, *Differentsial'nyye Uravneniya,* **8**, 1 (1972).
141. V. G. Romanov, *Nekotoryye obratnyye zadachi dlya uravneniy giperbolicheskogo tipa* (Some inverse problems for hyperbolic equations), Novosibirsk, Nauka Press, 1972.
142. V. G. Romanov, *Obratnyye zadachi dlya differentsial'nykh uravneniy* (Inverse problems for differential equations), Novosibirsk State University Press, 1973.
143. V. G. Romanov, The abstract inverse problem and questions regarding its posing, *Funktsional'nyy analiz,* **7**, 3 (1973).
144. T. I. Savelova and V. V. Tikhomirov, The solution of integral equations of the first kind of the convolution type in the multidimensional case, *Zhurnal vychislitel'noy matematiki i matematicheskoy fiziki,* **13**, 3 (1973).
145. T. I. Savelova, Solutions of the convolution type with inexactly given kernel by the regularization method, *Zhurnal vychislitel'noy matematiki i matematicheskoy fiziki,* **12**, 1 (1972).
146. T. I. Savelova, Application of a class of regularizing algorithms to the solution of integral equations of the first kind of the convolution type in a Banach space, *Zhurnal vychislitel'noy matematiki i matematicheskoy fiziki,* **14**, 2 (1974).
147. T. I. Savelova, Projection methods of solving linear ill-posed problems, *Zhurnal vychislitel'noy matematiki i matematicheskoy fiziki,* **14**, 4 (1974).

148. V. N. Strakhov, The solution of ill-posed problems of magnetometry and gravimetry in the form of integral equations of the convolution type, *Izvestiya Akad. Nauk SSSR,* Ser. Fizika Zemli, **4**, 5 (1967).
149. V. N. Strakhov, Numerical solution of ill-posed problems represented by integral equations of the convolution type (*Doklady Akad. Nauk SSSR*), **178**, 2 (1968).
150. V. N. Strakhov, Linear ill-posed problems in a Hilbert space, *Differentsial'nyye Uravneniya,* **6**, 8 (1970).
151. V. N. Strakhov, A method of successive approximations for linear equations in a Hilbert space, *Zhurnal vychislitel'noy matematiki i matematicheskoy fiziki,* **13**, 4 (1973).
152. V. N. Strakhov, The construction of approximate solutions, optimal with respect to order, of linear conditionally well-posed problems, *Differentsial'nyye Uravneniya,* **11**, 10 (1973).
153. V. N. Strakhov, The speed of convergence in the method of simple iteration, *Zhurnal vychislitel'noy matematiki i matematicheskoy fiziki,* **13**, 6 (1973).
154. V. P. Tanana, Approximate solution of operator equations of the first kind in locally convex spaces, *Izvestiya vuzov, matematika,* **9** (1973).
155. A. N. Tikhonov, The stability of inverse problems, *Doklady Akad. Nauk SSSR,* **39**, 5 (1943).
156. A. N. Tikhonov, The solution of ill-posed problems, *Doklady Akad. Nauk SSSR,* **151**, 3 (1963).
157. A. N. Tikhonov, Regularization of ill-posed problems, *Doklady Akad. Nauk SSSR,* **153**, 1 (1963).
158. A. N. Tikhonov, Stable methods of summing Fourier series, *Doklady Akad. Nauk SSSR,* **156**, 1 (1964).
159. A. N. Tikhonov, Solution of nonlinear integral equations, *Doklady Akad. Nauk SSSR,* **156**, 6 (1964).
160. A. N. Tikhonov and V. B. Glasko, Approximate solution of Fredholm integral equations of the first kind, *Zhurnal vychislitel'noy matematiki i matematicheskoy fiziki,* **4**, 3 (1964).

161. A. N. Tikhonov, Nonlinear equations of the first kind, *Doklady Akad. Nauk SSSR,* **161**, 5 (1965).
162. A. N. Tikhonov, V. Ya. Arsenin, L. A. Vladimirov, G. G. Doroshenko, and A. A. Dumova, The processing of apparatus spectra of gamma-quanta and fast neutrons measured with the aid of single-crystal scintillation spectrometers, *Izvestiya Akad. Nauk SSSR,* Ser. Fizicheskaya, **XXIX**, 5 (1965).
163. A. N. Tikhonov, Methods of regularizing optimal control problems, *Doklady Akad. Nauk SSSR,* **162**, 4 (1965).
164. A. N. Tikhonov, V. Ya. Arsenin, A. A. Dumova, L. V. Mayorov, and V. I. Mostovoy, A new method of restoring true spectra, *Atomnaya energiya,* **18**, 6 (1965).
165. A. N. Tikhonov and V. B. Glasko, Application of regularization methods in nonlinear problems, *Zhurnal vychislitel'noy matematiki i matematicheskoy fiziki,* **5**, 3 (1965).
166. A. N. Tikhonov, Ill-posed problems of linear algebra and a stable method of solving them, *Doklady Akad. Nauk SSSR,* **163**, 6 (1965).
167. A. N. Tikhonov, The stability of algorithms for solving singular systems of linear algebraic equations, *Zhurnal vychislitel'noy matematiki i matematicheskoy fiziki,* **5**, 4(1965).
168. A. N. Tikhonov, Ill-posed optimal planning problems and stable methods for solving them, *Doklady Akad. Nauk SSSR,* **164**, 3 (1965).
169. A. N. Tikhonov, Methods for solving ill-posed problems, Theses of addresses at the International Congress of Mathematicians, Moscow, 1966.
170. A. N. Tikhonov, Ill-posed optimal planning problems, *Zhurnal vychislitel'noy matematiki i matematicheskoy fiziki,* **6**, 1 (1966).
171. A. N. Tikhonov, Stability of the problem of minimizing functionals, *Zhurnal vychislitel'noy matematiki i matematicheskoy fiziki,* **6**, 4 (1966).
172. A. N. Tikhonov, V. Ya. Galkin, and P. N. Zaikin, Direct methods of solving optimal control problems, *Zhurnal*

vychislitel'noy matematiki i matematicheskoy fiziki, 7, 2 (1967).

173. A. N. Tikhonov and V. B. Glasko, Methods of determining the temperature of the surface of bodies, *Zhurnal vychislitel'noy matematiki i matematicheskoy fiziki,* 7, 4 (1967).
174. A. N. Tikhonov, Ill-posed problems, in the collection *Vychislitel'naya matematika i programmirovaniye* (Computational mathematics and programming), **VIII**, Moscow State University Press, 1967.
175. A. N. Tikhonov, V. V. Alikayev, V. Ya. Arsenin, and A. A. Dumova, Determination of the distribution function of the electrons of a plasma from the spectrum of Bremsstrahlungen, *Zhurnal eksperimental'noy i teoreticheskoy fiziki,* **55**, 5 (1968).
176. A. N. Tikhonov, V. G. Shevchenko, V. Ya. Galkin, P. N. Zaikin, B. I. Goryachev, B. S. Ishkhanov, and I. M. Kapitonov, A system of continuous automatic processing of the results of an experiment from investigation of the sections of photonuclear reactions, in the collection *Vychislitel'naya matematika i programmirovaniye* (Computational mathematics and programming), **XIV**, Moscow State University Press, 1970.
177. A. N. Tikhonov and V. I. Dmitriyev, Methods of solving the inverse antenna theory problem, in the collection *Vychislitel'naya matematika i programmirovaniye* (Computational mathematics and programming), **XIII**, Moscow State University Press (1969).
178. A. N. Tikhonov, V. G. Karmanov, and T. L. Rudneva, Stability of linear programming problems, in the collection *Vychislitel'naya matematika i programmirovaniye* (Computational mathematics and programming), **XII**, Moscow State University Press, 1969.
179. V. F. Turchin, Solution of a Fredholm equation of the first kind in the statistical aggregate of smooth functions, *Zhurnal vychislitel'noy matematiki i matematicheskoy fiziki,* 7, 6 (1967).

180. V. F. Turchin, Choice of the aggregate of smooth functions in the solution of the inverse problem, *Zhurnal vychislitel'noy matematiki i matematicheskoy fiziki,* **8**, 1 (1968).
181. V. F. Turchin, V. P. Kozlov, and M. S. Malkevich, Use of the methods of mathematical statistics to solve ill-posed problems, *Uspekhi fizicheskikh nauk,* **102**, 3 (1970).
182. L. S. Frank and L. A. Chudov, Difference methods for solving an ill-posed Cauchy problem, in the Collection *Vychislitel'naya matematika i programmirovaniye* (Computational mathematics and programming), **XIV**, Moscow State University Press, 1965.
183. L. S. Frank, Difference methods for solving the Cauchy problem for first-order ill-posed systems, *Vestnik,* Moscow State University, Matematika, **1** (1966).
184. L. A. Khalfin and V. N. Sudakov, A statistical approach to correctness of posing of the problems of mathematical physics, *Doklady Akad. Nauk SSSR,* **157**, 5 (1964).
185. Yu. I. Khudak, Regularization of the solutions of integral equations of the first kind, *Zhurnal vychislitel'noy matematiki i matematicheskoy fiziki,* **6**, 4 (1966).
186. Yu. I. Khudak, Convergence of a family of regularizing operators, *Zhurnal vychislitel'noy matematiki i matematicheskoy fiziki,* **12**, 2 (1972).
187. Yu. I. Khudak, Convergence of regularizing algorithms, *Zhurnal vychislitel'noy matematiki i matematicheskoy fiziki,* **11**, 1 (1971).
188. L. A. Chudov, Difference schemes and ill-posed problems for partial differential equations, in the collection *Vychislitel'naya matematika i programmirovaniye,* **VIII**, Moscow State University Press, 1967.
189. R. Arcangeli, Pseudosolution de l'équation $Ax = y$, *Comptes Ren. Acad. Sci.,* **263**, 8 (1966).
190. R. Bellman, R. Kalaba, and J. Lochett, Dynamic programming and ill-conditioned linear systems, *J. Math. Anal. and Appl.*, 10 (1965).
191. A. Bensonssan and P. Kenneth, Sur l'analogue entre des

méthodes de régularisation et de pénalisation, *Rev. franc. inform. et rech. oper.* **2**, 13 (1968).

192. H. S. Cavayan and G. G. Belford, On computing a stable least squares solution to the inverse problem for a planar Newtonian potential, SIAM, *J. Appl. Math.*, **20**, 1 (1971).
193. S. Cruceanu, Régularisation pour les problèmes à opérateurs monotones et la méthode de Galerkine, *Comment. Math. Univ. Carol.*, **12**, 1 (1971).
194. J. Douglas, A numerical method for analytic continuation, Boundary Problems Different. Equat. Univ. Wisconsin Press, Madison, 1960.
195. J. Douglas, Mathematical programming and integral equations, Sympos. Numerical Treatm. Ordinary Differential Equations, Integral and Integro-different. Equat. Birkhauser, 1960.
196. J. Douglas, Approximate continuation of harmonic and parabolic functions, *Numerical Solution of Partial Differential Equations*, Acad. Press, New York, 1966.
197. J. Douglas and T. Callie, An approximate solution of an improper boundary value problem, *Duke Math. J.*, **26**, 3 (1959).
198. D. W. Fox and C. Pucci, The Dirichlet problem for the wave equation, *Annali di Math.*, **46** (1958).
199. G. Fichera, Sul Concetto di problemi "ben posti" per una equazione differentiale, *Rendiconti di Matematica e delle sue applicazioni,* **19**, 1 (1960).
200. M. Furi and A. Vignoli, On the regularization of nonlinear ill-posed problem in Banach spaces, *J. Optimiz. Theory and Appl.*, **4**, 3 (1969).
201. J. Hadamard, Sur les problèmes aux dérivées partielles et leur signification physique, Bull. Princeton Univ., **13** (1902).
202. J. Hadamard, *Le problème de Cauchy et les équations aux dérivées partielles linéaires hyperboliques*, Hermann, Paris, 1932.
203. F. John, Numerical solution of the equation of heat conduction for preceding times, *Ann. Mat. Pura ed Appli.*, **40**, (1955).

204. F. John, A note on "improper" problems in partial differential equations, *Comm. Pure and Appl. Math.,* **8** (1955).
205. F. John, Continuous dependence on data for solutions with a prescribed bound, *Comm. Pure and Appl. Math.,* **13**, 4 (1960).
206. F. John, Numerical solution of problems which are not well-posed in the sense of Hadamard, *Symposium on Numerical Treatment of Partial Differential Equations with Real Characterictics,* Rome, 1959.
207. K. Marton and L. Varga, Regularization of certain operator equations by filters, *Stud. Sci. Math. Hung.*, **6**, 3 (1971); **6**, 4 (1971).
208. I. P. Nedyalkov, An approach in the theory of incorrect problems, *Dokl. Bolg. AN,* **23** (1970).
209. D. J. Newman, Numerical method for solution of an elliptic Cauchy problem, *J. Math. and Phys.,* **39**, 1 (1960).
210. L. E. Payne, Bounds in the Cauchy problem for the Laplace equation, *Arch. Rational. Mech. Anal.,* **5**, 1 (1960).
211. D. L. Phillips, A technique for the numerical solution of certain integral equations of the first kind, *J. Assoc. Comput. Mach.,* **9**, 1 (1962).
212. C. Pucci, Sui problemi Cauchy non "ben posti", *Atti Acad. Naz. Lincei,* Rend. Cl. Sc. fis. mat. et natur., **18**, 5 (1955).
213. C. Pucci, Discussione del problema di Cauchy pur le equazioni di tipo ellittico, *Ann. mat. pur ed appl.,* **46** (1958).
214. C. Pucci, On the improperly posed Cauchy problems for parabolic equations, *Symposium on Numerical Treatment of Partial Differential Equations with Real Characteristics,* Rome, 1959.
215. J. Replogle and B. D. Holcomb, The use of mathematical programming for solving singular and poorly conditioned systems of equations, *J. Math. Anal.,* **20**, 2 (1967).
216. G. Ribière, Régularisation des opérateurs, *Rev. frans. inform. et rech. oper.,* **1**, 11 (1967).

217. C. B. Shaw, Improvement of the resolution of an instrument by numerical solution of an integral equation., *J. Math. Anal. and Appl.*, **37**, 1 (1972).
218. O. N. Strand and E. R. Westwater, Statistical estimation of the numerical solution of Fredholm integral equations of the first kind, *J. Assoc. Comput. Math.*, **15**, 1 (1968).
219. S. Twomey, On the numerical solution of Fredholm integral equations of the first kind by the inversion of the linear system produced by quadrature, *J. Assoc. Comput. Mach.*, **10**, 1 (1963).
220. S. Twomey, The application of numerical filtering to the solution of integral equations encountered in indirect sensing measurements, *J. Franklin Inst.*, 1965.
221. N. Wiener, *Extrapolation, Interpolation and Smoothing of Stationary Time Series with Engineering Applications*, New York, Wiley, 1950.

SUBJECT INDEX